Energy Efficient Vehicles 2014

Bernard Bäker, Lutz Morawietz (Eds.)
and 56 Co-Authors

TUDpress

Energy Efficient Vehicles 2014

Visions, Trends and Solutions for Energy Efficient Vehicle Systems

Prof. Dr.-Ing. Bernard Bäker
Dipl.-Ing. Lutz Morawietz (Eds.)
and 56 Co-Authors

TUD*press*
2014

Bibliografische Information der Deutschen Nationalbibliothek
Die Deutsche Nationalbibliothek verzeichnet diese Publikation in der Deutschen Nationalbibliografie; detaillierte bibliografische Daten sind im Internet über http://dnb.d-nb.de abrufbar.

Bibliographic information published by the Deutsche Nationalbibliothek
The Deutsche Nationalbibliothek lists this publication in the Deutsche Nationalbibliografie; detailed bibliographic data are available in the Internet at http://dnb.d-nb.de.

ISBN 978-3-944331-65-2

Verlag der Wissenschaften GmbH
Bergstr. 70 | D-01069 Dresden
Tel.: 0351/47 96 97 20 | Fax: 0351/47 96 08 19
http://www.tudpress.de

Printed in Germany.

Preface

The fulfillment of a given automotive transportation task in a most energy efficient way while keeping a high level of security and comfort is still one of the main challenges in the development of modern vehicle systems nowadays and in the future. The corresponding objectives have to be addressed with a close interaction between different scientific competency areas and different development domains like the electric and electronic engineering, the computer science, the mechanical engineering, the vehicle mechatronics, mathematics supported by the automotive industry in combination with engineering service providers, system suppliers and technical universities.

This book presents the proceedings of the 3rd International Conference on Energy Efficient Vehicles (ICEEV 2014) which took place from 24th June -25th June 2014 in Dresden, Germany.

The conference follows the successful "Energy Efficient Vehicle Conference" and intends to provide a platform for the interaction between researchers, manufacturers, developers and managers of all the above mentioned scientific domains and therewith to build up new visions, trends and solutions for energy efficient vehicle solutions.

The exciting conference program for the ICEEV 2014 contained contributions from energy efficient power net solutions to optimized onboard operating strategies and further more.

We would like to thank all the authors and speakers. Only with their help the realization of this book was possible. We would also like to thank Mrs. Bock and Mrs. Wrenczycki from CMD Congress Management Dresden GmbH, Dresden and Mrs. Schüßler from TUDpress Verlag, Dresden for the excellent cooperation.

Sincerely yours,

Bernard Bäker   Lutz Morawietz

Contents

Keynote

1 Vehicle Aerodynamics and Thermal Management in Future Hybrid and Electrical Vehicles

Prof. Dr.-Ing. **Lennart Löfdahl**, Applied Mechanics, Chalmers University of Technology, Gothenburg, Sweden

Abstract

After more than 100 years of combustion engine use for vehicle propulsion, a most relevant and slightly provocative question is: why is there so much interest now in hybrid vehicles development? The answer is multi-faceted, however there are basically three main arguments for the revitalized interest: less CO2-emissions and significantly lower fuel consumptions; improved driving dynamics and comfort through a second energy converter which for a limited time can deliver increased driving torque and/or acceleration; and last but certainly not least, the development of hybrid vehicles act as an intermediate and necessary stage in the on-going development towards fully electrical vehicles.

Development of hybrid vehicles faces many new technical challenges; a number of components in the drive-train need more space and add weight to the vehicle, the battery is a particularly demanding device. Furthermore, customers require that a hybrid vehicle, despite its increased complexity, should have the same performance, reliability and cost profile as a conventional one. This certainly puts heavy demands on the vehicle development. Improved battery technology, better electric motors and power electronics are necessary, requiring optimization of vehicle aerodynamics and thermal management.

Three concepts are common in discussions on hybrid and electrical vehicles: energy management, road vehicle aerodynamics, and thermal management. Many times this terminology is ambiguous and may create miss-understandings and even incorrect interpretations. In the paper these concepts are defined, sorted out and discussed.

Main focus of the paper is on the use of vehicle aerodynamics and thermal management on hybrid and electrical vehicles. A brief history of hybrid vehicles, starting with one of the first hybrid cars that was built back in 1899 by Ferdinand Porsche, is given as an introduction. This vehicle was the System Lohner-Porsche Mixte, and used a gasoline engine to supply power to an electric motor which in turn drove the front wheels. From the historical archive, milestones are elucidated and analysed from today's environmental perspective. The historical part finishes with a discussion based on the ever-increasing environmental awareness; and that the Toyota Prius (for example) may be in for some stiff competition.

To increase the relevance of motor sport for production car design a number of new regulations were introduced for the 2014 season. Specifically for F1 a massive fuel consumption challenge, 100kg/h fuel flow limit and a maximum fuel allocation of 100 kg, was introduced. To meet these conditions new power units were developed due to a fundamentally different rulebook. This means the return of turbo-charging and a massive increase in potency of the hybrid system. Hence, the F1 cars for 2014 may

be categorized as hybrid electrical vehicles since they combine a conventional internal combustion engine with an electrical propulsion system, and may be considered as the most energy-efficient vehicles of today. In this context, the energy recovery systems such as KERS (Kinetic Energy Recovery System), MGU (Motor Generator Unit) and recuperative braking are all used to the limit and in the paper these devices are reviewed and discussed.
In a similar way as for passenger and race cars there is a potential of introducing hybrids for medium and heavy trucks. Benefits from a second energy converter in the next generation of trucks are illustrated, and possibilities opened by hybrid propulsion in new vehicle architectures are summarized and discussed.
Lithium-ion batteries show strong dependence on their operating temperatures, too low temperature will limit the energy efficiency, while a too high will accelerate the aging of the battery. Hence, the maintenance of good operating temperature (climate control) of the battery storage is of utmost importance for the functioning of hybrid and electrical vehicles. A number of important factors, such as the temperature of the ambient-air, engine, electrical motors and load cycles of the vehicle interact and must be controlled in order to optimize drive range and battery aging. Advanced models for the description of the different systems and their interaction are needed.

1. Introduction

After more than 100 years use of combustion engines for propulsion, a relevant but slightly provocative question to pose is why hybrid electrical vehicles (HEV) are so interesting to develop. The answer is multifaceted and there are basically three main arguments for the interest: Less CO2-emissions and significantly lower fuel consumptions, improved driving dynamics and comfort through a second energy converter which (for a limited time) can deliver increased driving torque as well as acceleration. Last but certainly not least, the development of HEVs act as an intermediate and necessary step in the on-going development towards fully electrical vehicles.
Development of HEVs faces many technical challenges, a number of components in the drive train need more space and add weight to the vehicle; especially the battery is a demanding device. Furthermore, customers require that a HEV, in spite of its increased complexity, should have the same performance, reliability and cost profile as a conventional vehicle and this places high demands on the development process. Improved battery technologies where increased capacity and lifetime are key issues, but also better electric motors and power electronics are needed.
The focus of this paper is to highlight the importance of Vehicle Aerodynamics and Thermal Management in the future HEVs growth; however, the intention is also to demonstrate the possibilities of "carry over" knowledge between different vehicle categories.
The paper starts with a discussion on the terminology used; concepts like Road Vehicle Aerodynamics, Energy-, and Thermal Management are discussed and defined. Some comments are given on the development history of hybrid vehicles and a more than hundred years old advanced HEV is highlighted. The most energy efficient vehicles of today are, through changes in the regulations toward more "green racing", found in the racing world and some features of the 2014 Formula One (F1) HEV cars are illustrated from a HEV perspective. Moreover, opportunities with hybrid

based drive train in heavy vehicles are discussed briefly, and a few remarks are made on the so important battery climate control issue.

2. Terminology & brief history

Three concepts are commonly used in discussions on HEV; Energy management, Road Vehicle Aerodynamics and Thermal Management. Many different definitions exist, see for instance Hucho [1]; however, the terminology is often mixed and used differently in different contexts so this might create misunderstandings and even erroneous interpretations. An attempt is made; based on fundamental reasoning from text-books in fluid- and thermo-dynamics, see e.g. Crowe et.al. [2], Panton [3], Bachelor [4] and Eckert and Drake [5], to sort out the three concepts.

Energy management is a very general concept and may be defined as: A continuous planning process utilized to accomplish the most efficient use of energy.
Slightly more restricted is *Thermal Management* and may be defined as: A process of adjusting the operating temperature of a certain device by applying heat peaks and optimized air flows. Temperatures, pressures and flows of the fluids are needed to be controlled in components and in the power-train to an optimum.
A fundamentally based definition of *Vehicle Aerodynamics* is to consider it to be studies of the aerodynamics of road vehicles with the main objective to reduce drag, emissions, wind noise, and, for certain applications, to produce down-force to improve traction and cornering abilities.
In a grey zone between Vehicle Aerodynamics and Thermal Management, with a foot in both camps, the so-called *Cooling-Drag* is found. For HEVs this area will in the future be of increased importance, since today there is a clear trend towards increased operating temperatures and more optimized heat transfer processes implying that the contribution from cooling-drag to the total drag increases.

3. A Brief History of Hybrid vehicles

The history of HEV spans for more than a century and in recent years there has been intense development. A common thread running through the history of HEV is that the authorities, especially in USA, have tried to support the development by tax incentives. For instance, in the 1960s, the US Congress introduced legislation aiming to encourage a larger use of HEVs to reduce air pollution. Over the next 25 years, the vehicle manufacturer (OEM) spent much money on research and development of HEVs; however, no vehicles were produced which were competitive to gasoline vehicles on neither price nor performance. It was not until around 2000 when the Toyota Prius and the Honda Insight were launched that a viable alternative to gas powered vehicles was available on the market. To cover the on-going development of HEVs is a comprehensive task and it is out of the scope for the present paper; however, an interesting overview of the early history will be illuminating.
One of the very first HEVs was designed by Ferdinand Porsche late in the nineteenth century. This vehicle is shown in figure 1, and it was named "Semper Vivus", which is Latin for "always alive."
Based on his race experience, Porsche took a pair of single-cylinder De Dion-Bouton engines; each producing (2.5 hp), to served two functions: Spin a generator which

provide power to the wheel hub motors and batteries; and to drive a water pump to provide coolant to the engines. The propulsion came exclusively from the hub-mounted electric motors, and no mechanical connection existed between the engines and the front drive wheels. As with many modern HEVs, the "Semper Vivus" was a heavy vehicle at 1,200 kilograms.

Figure 1: *Ferdinand Porsche's 1901 "Semper Vivus". (Photos courtesy Porsche AG.)*

By the 1901 Paris Motor Show, Porsche had put the "Semper Vivus" behind him and accepted that it was not a commercially viable product. He was working on an improved version, using a single 5.5-liter Austrian Daimler engine (25 hp) mounted in the front of the vehicle. A large generator was positioned beneath the driver's bench seat, and under normal operation, the engine and the generator was run at a constant speed to supply, via the battery, power for the hub-mounted motors. This version was called the "Lohner-Porsche Mixte" and is shown in figure 2.

Figure 2: *Ferdinand Porsche at the wheel of a 1903 "Lohner-Porsche Mixte". (Photos courtesy Porsche AG.)*

The argument against electric power at the turn of the century was exactly the same as today: Limited range and a lack of charging infrastructure make the electric vehicles impractical for many potential buyers. In spite of this, the production of the "Lohner-Porsche Mixte" continued; however, with numerous changes and improvements until 1915. The vehicle was never a commercial success since, due to several sources, it was simply too expensive.

Although it is a quantum leap between the "Lohner-Porsche Mixte and today's most energy efficient vehicles the number of challenges for the OEMs in terms of underperforming batteries, too heavy vehicles and difficulties to motivate a higher price for the customers seems to be the same.

4. "The engine is no more, long live the Power Unit"

The title of this paragraph is a quotation from the Mercedes-Benz Petronas description of their F1 engines for the 2014 season. Due to far-reaching changes in the regulations, the F1 cars of today, an example is shown in figure 3, are probably the most energy efficient vehicles in the world.

Figure 3: *Mercedes-Benz Petronas 2014, one of the worlds most energy efficient vehicles.*

Through the introduction of Kinetic Energy Recovery System Hybrid power in 2009, the idea of having only the engine as a standalone source of propulsion was dispatched to the F1 history. The 2014 regulation is far-reaching; a 1.6 litre, turbocharged V6 configuration together with an integrated Hybrid Energy Recovery System (ERS) form the power unit. So in principle the power unit of 2014 comprises six different sub-systems: the internal combustion engine (ICE), Motor Generator Unit-Heat (MGU-H), Motor Generator Unit-Kinetic (MGU-K), Energy Store (ES), Turbo-charger and the Control-Electronics. These different units are shown in figure 4 in the Renault F1 outline.
For the ICE the regulation demands that the crankshaft speed is reduced to 15000 rpm and that the number of cylinders is reduced to six, thus, the internal engine friction decrease and the total efficiency of the power unit increases. This down-sizing is a key issue which will be introduced in many HEV concepts and will change the ICE structure significantly. Also interesting is that energy is recovered and deployed

to the rear axle via the two MGUs. As compared to previous years, the 2014 regulation requires that significantly larger amounts of energies are recovered; the MGU-K has twice as large maximum power as previously and the MGU-H is an electrical machine connected to the turbocharger with no energy limit, so this unit can deliver a substantial amount of extra energy to the drive train.

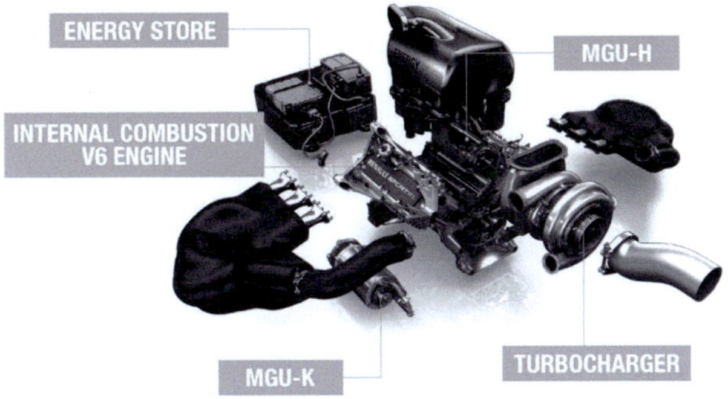

Figure 4: *Systems in the Renault power unit outline for 2014.*

In the energy storage (ES), the energy harvested from the MGUs is stored for deployment back into the drive train. As with all batteries there is always a weight problem, and to overcome this, the ES is weight regulated. A battery package is shown in figure 5.

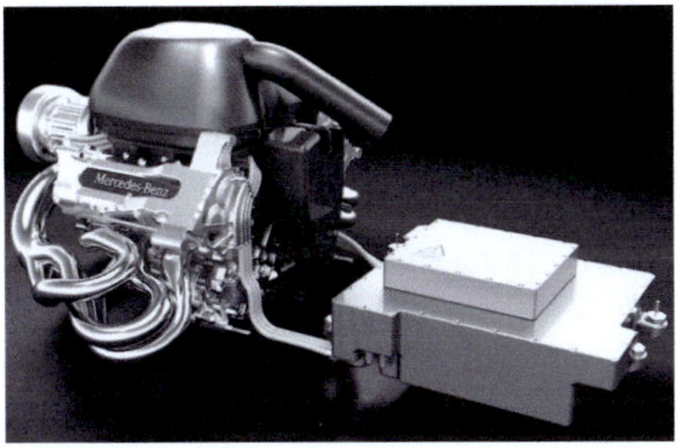

Figure 5: *Mercedes-Benz Petronas power unit outline with battery package.*

The fact that a full F1-race in the 2014 season must be carried out on 100 kg fuel (150 kg was typical for a race in 2013) and at a maximum flow rate of 100 kg/h is a tough challenge. To cope with this, not only energy recovery is a necessity, but also the total fluid dynamic drag of the vehicle must be kept at a minimum level and this implies that flow drag issues will be in focus much more than it has been during previous years. It might be plausible to expect that the external design of F1 vehicles might be very different in the years to come.
All power unit system introduced in 2014 means that more equipment have a tendency to be hot and need efficient cooling. For instance, an intercooler for the

charge air to the turbocharger is a necessity so the classical HEV problem of more and heavy equipment has also been faced by the F1 designers. For a race car this has to be an extreme compromise between packaging, Vehicle Aerodynamic and a Thermal Management. This is a real challenge and in the early part of the 2014 season this difficulty was quite obvious. As is shown in figure 6 heat problems was not a unique event and some of the large teams had a very limited test periods for the season 2014. However, many new efficient radiator systems have been launched recently, but still, as compared to previous years the F1 of 2014 are extremely sensitive to heat problems. Positioning of the ES has turned out to be a critical issue, and it is interesting to note that currently only Ferrari has shown a battery cooling system.

Figure 6: *A Red Bull with heating problem.*

The F1-regulation of 2014 introduced many Vehicle Aerodynamics and Thermal Management issues, and this also points out the direction of development for standard vehicles. Minimal fuel consumption and optimized thermal efficiency under heavy packing restrictions are together with weight limitations challenges for all future HEVs.

To underline the importance of fluid drag on HEVs, Wiedemann et.al [2] has recently compared a reduction of the aerodynamic drag to battery cost (range of operation) for different drag coefficients and weights of the vehicle. It was found that reduced flow drag is the most efficient way to increase the operating range (reduce battery cost) of a HEV.

5. Hybrids in medium and heavy trucks

For commercial vehicles the strict legal requirements demands most energy efficient vehicles. Today complex systems on the ICE for emission reduction are used together with elaborate exhaust after-treatment concepts. A good summary of the systems used in modern trucks can be found in Töpfel et.al. [3]. One conclusion made in this work is that not only the power-train has to be energy efficient, but due to stricter legal requirements the complete vehicle must be put under the microscope for further CO_2 savings. Using Sankey-diagram, se figure 7, Töpfel et al. [3] compared a conventional and a future power-train and conclude that increased

number of participating sub-systems must be included in the holistic view and this is necessary for improved energy efficiency.
Looking at figure 7 it can be noted that many features discussed in previous section on race cars also may find their applications on commercial vehicles. Energy optimized recovery of waste heat energy, systems for kinetic energy recuperation, energy storage and electrical auxiliary systems (coolant, oil, compressed air et.) will be needed. A common denominator of several of these actions is that they all have a tendency to increase the working temperature of the power-train, and hence, they all put heavy demands on the cooling system and Thermal Management. A simple solution might be to increase the cooling capacity; however, this far from a straight-forward task since it will require more cooling fluid and optimized air intakes and outlets. All this needs space and increases weight.

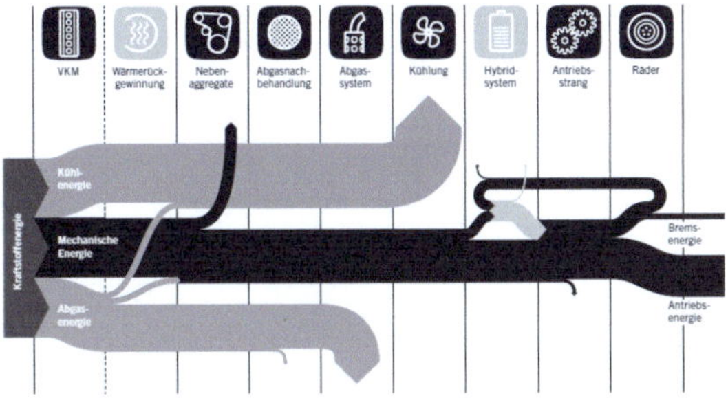

Figure 7: *A Sankey-diagram for a future power train unit. Töpel et.al. [3]*

New vehicle architectures are briefly mentioned in the paper by Töpel et.al. [3] and three examples are shown in figure 8. Although these vehicles must be considered as simple general outlines they point clearly to the possibilities of moving the second energy converter to different positions on the truck. For future medium- and heavy-trucks this is a great possibility since the range of operation may be increased and the down-sizing could be significantly more efficient.

6. Climate control of the battery storage

It is a well-known fact that the available power and capacity of lithium-ion-batteries shows a strong dependence on the operating temperature. On the one hand, low temperature limits the driving range, especially under winter conditions; while on the other hand, a too high temperature accelerates the ageing and influences the thermal stability of the battery materials. As is shown in figure 9, a HEV consists, in principle, of at least five different thermal systems with relatively known operating conditions and temperatures.
For warming up the battery package at low ambient air temperature, the relatively small amount of waste heat from the electric powertrain must be integrated with the heat flux from the combustion engine, and tentatively also from the exhaust system. The cooling from the A/C system will be essential for the battery package at high outdoor temperatures. In addition, the battery itself produces heat when loaded so for

battery climate control the driving cycle of the vehicle must be considered. Taken all together there are most stringent requirements on the control system for the temperature in the battery compartment since all related system components must be modelled under transient conditions.

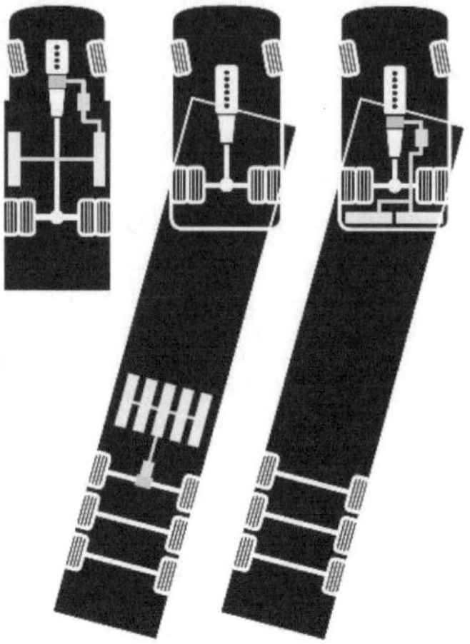

Figure 8: *Three different HEV drive train outlines.*
From Töpel et.al. [3]

Hopp et.al. [8] has studied thermal simulation of high-performance battery systems by analysing the cooling system of a HEV sports car. Their simulation model of the full vehicle is divided into individual simulation sub-models which are all coupled. The sub-models include the thermal-electrical battery model, fluid cooling system model, refrigerant circuit model and the power train model. Particularly high-load cycles such as race-tracks are modelled and analysed.
Since the work of Hopp et.al. [8] was presented, a number of optimization studies have been published; however, most of them are not capable of physically modelling the different sub-systems under transient load conditions and in order to extend the driving range of HEVs these models must be developed further.

7. Conclusion and Future Outlook

Some on-going development of HEV vehicles has been reviewed and discussed; specifically the use and potential of Vehicle Aerodynamics and Thermal Management in the development process have been illustrated.
As an epilogue it may be suitable to highlight, selected from numerous impressive concepts recently presented by the OEMs, the Audi e-tron Spyder's bi-turbo diesel, shown in figure 10. This vehicle has breath-taking performance, it is capable of making 0-62mph in 4.4 seconds, has a top speed of 155mph, combined economy of 128.4mpg, CO_2 emissions of just 59g/km and a total range of approximately 1000 km. The plug-in hybrid combines a 300 PS, bi-turbo version of a 3.0-litre TDI engine

and two electric motors. The total weight of around 1,450 kilograms is low, and helps to improve performance as well as economy.

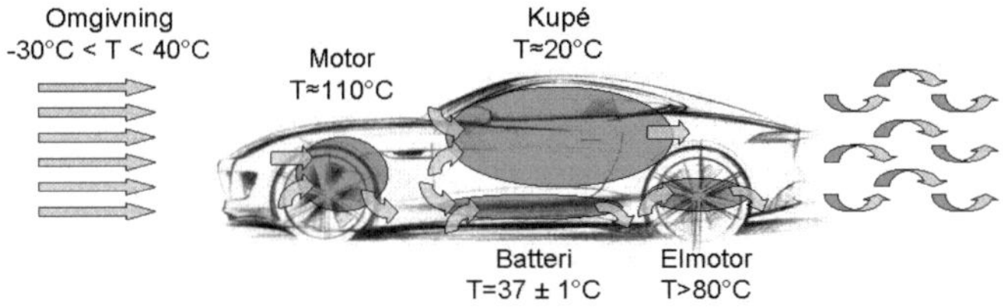

Figure 9: *Principle sketch of the different thermal systems in a hybrid vehicle.*

The body structure is based on Audi Space Frame technology with elements like the bonnet and numerous aerodynamic components made out of carbon to keep weight low. This low weight, short wheelbase and perfect 50:50 weight distribution for dynamic handling gives the Audi e-tron Spyder go-kart like drivability.

This vehicle sacrifices nothing in terms of performance and it has an incredible potential for low fuel consumption. From a Vehicle Aerodynamics and Thermal Management perspective it is, of course, the low drag design of the body that is impressive, also the aggressive and efficient wheel-rims, and air-intakes at the front.
In conclusion it can be stated that the Audi e-tron Spyder is one out of many HEV-concepts showing that no real technical limits exists in the design of HEVs.

Figure 10: *Audi e-tron Spyder.*

References

[1] Hucho W-H.; "Aerodynamik des Automobils", ATZ/MTZ-Fachbuch, Vieweg Teubner, ISBN 978-3-528-03959-2, 2005.

[2] Crowe C., Roberson J. & Elger D.; "Engineering Fluid Mechanics", 7th ed., John Wiley & Sons, New York, ISBN 0-471-38482-8.

[3] Panton R.; "Incompressible Flow", 3rd ed., John Wiley & Sons, Hoboken, New Jersey, ISBN-13 978-0-471-26122-3.

[4] Bachelor G.; "An Introduction to Fluid Dynamics", Cambridge University Press, Cambridge, ISBN 0-512-04118.

[5] Eckert E. & Drake R.; "Analysis of Heat and Mass Transfer", McGraw-Hill Book Company, USA.

[6] Wiedemann J., Wiesebrock A., & Heidorn H., (2013); "Kann man mit dem Luftwiderstandsbeiwert die Batteriekosten senken? Energieeffiziente Antriebstechnologien: Hybridisierung – Downsizing – Software und IT", ISBN 978-3-658-00789-8, Springer Fachmedien, Wiesbaden.

[7] Töpfer T., Henning L., Eckert P., & Seebode J., (2013); Energiemanagement für Nutzfahrzeuge der nächsten Generation, Energieeffiziente Antriebstechnologien Hybridisierung – Downsizing – Software und IT", ISBN 978-3-658-00789-8, Springer Fachmedien, Wiesbaden.

[8] Hopp H., Wiedemann J., Lemke T., & Widdecke N. (2013); Thermal simulation of high-performance battery systems, FKFS Symposium, Stuttgart.

2 Risks and Solutions for Modern Energy Harvesting

Prof. Dr. Dr. **Gerhard Hettich**, EAST-Consulting, Dietenhofen, Germany

Abstract

Because to good reasons current systems of energy harvesting and distribution are not only carefully examined by politics and associations, but they are as well assessed in terms of their emissions, their degree of environmental friendliness and their resource requirements. A comparable critical examination of alternative solutions, like the one of electric traction, is very often subordinated to individual hopes, like the desire to control changes in climate. Consequently, we often deliberately fail to notice the fact that complex systems are not that easily controlled – this is a task that could not be accomplished by just adjusting a few setscrews. This fact and the industrial challenge of mass production and disposal of systems are vital aspects to consider when you would like to come to logical alternative conclusions.

Kurzfassung

Aus guten Gründen werden die heutigen Energienutzungssysteme von Politik und Verbänden kritisch unter die Lupe genommen und bezüglich ihrer Emissionen, Umweltverträglichkeit, aber auch ihres Ressourcenbedarfs, bewertet. Eine vergleichbare kritische Prüfung von alternativen Lösungen, wie zum Beispiel die der Elektrotraktion, wird aber häufig ideellen Zielen wie der Hoffnung, Klimaänderungen beeinflussen zu können untergeordnet. Dadurch übersieht man gerne, dass komplexe Systeme eben nicht durch einige wenige Stellschrauben zu beherrschen sind. Dies und die industrielle Herausforderung für die Massenproduktion und Entsorgung von Systemen sind wichtige Aspekte will man zu schlüssigen alternativen Konzepten kommen.

1. Einleitung und Problemstellung

Seit Leben auf unserem Planeten entstanden ist, hat dieses Leben den Planeten verändert. Manche der Änderungen, wie die Photosynthese, haben unsere Existenz erst möglich gemacht. Andere wie die Entstehung der Kohle – und Öllagerstätten ermöglichten uns deren Nutzung und den Aufbau einer industrialisierten Welt.
Nun mehren sich aber die Stimmen, die vor einer weiteren ungebremsten Ressourcennutzung auf unserer Erde warnen. Die erste Studie, die weltweit Aufmerksamkeit erregte war 1972 „Grenzen des Wachstums“ (1) von Dennis L. Meadows et al. Diese Studie schätzte Verbräuche gegen vorhandene Ressourcen ab und kam zu teilweise erschreckenden Aussagen was die Reichweite einzelner industrierelevanter Grundstoffe betraf. Obwohl die Voraussagen nicht in der prognostizierten Schärfe eintrafen, war es doch der Startschuss für eine bis heute anhaltende Diskussion.

Ein wichtiger Punkt ist dabei der weiter zunehmende Energieverbrauch, getrieben einerseits durch die zunehmende Technisierung und Globalisierung und andererseits durch die weiterhin ungebremste Urbanisierung. In Europa betrug der pro Kopf Verbrauch 2013 an elektrischer Energie 6200 kWh/a oder im Mittel eine Leistung von 0,7 kW pro Person.
Gedeckt wird der zunehmende elektrische Energiebedarf heute noch im Wesentlichen durch fossile Energieträger und Kernkraftwerke. Beide Arten der Energieerzeugung sind aber zumindest in Deutschland heftig umstritten.
Wir müssen uns deshalb einer Herausforderung stellen die hauptsächlich folgende Punkte betrifft:

- Verbrauch natürlicher Ressourcen
- Umweltbelastung durch Schadstoffe
- Klimabelastung

An der Bewältigung dieser Herausforderung sind nun 3 Gruppen beteiligt:

- die Politik, die handeln kann, aber Rücksicht auf andere Staaten und natürlich auf den Wähler nehmen muß

- die Erzeuger und Verbraucher, die Umsatz- und Konsum als Maßstab für Erfolg und Zivilisation betrachten

- die Experten mit all ihren menschlichen und wissenschaftlichen Schwächen
  In dieser Gemengelage spielt sich die Diskussion über die richtigen Maßnahmen für die Gestaltung unserer Zukunft ab. Daß dabei unzählige rationale, emotionale und teilweise esoterische Meinungen öffentlich werden, muß einen in einer vernetzten Welt nicht wundern.

Allein das Thema Klimawandel bringt bei Google 4 300 000 Zitate, eine Meinungsflut die kaum mehr einen nützlichen Beitrag zur Entscheidungsfindung liefern kann.

2. Das Problem mit den Vorhersagen

Die Schätzungen für die Reichweite der globalen Ölvorräte liegen seit einigen Jahrzehnten bei konstant ca. 40 Jahren. Solche Schätzungen beruhen jedoch immer auf den verfügbaren Fördermethoden, ändern sich diese, ändert sich auch der Ölvorrat. Ein Beispiel dafür ist das gut dokumentierte Kern-River-Ölfeld bei Bakersfield CA. Hier wurden generell die Schätzungen der noch zu fördernden Ölmenge durch die tatsächliche Förderung überboten. Grund hierfür waren neue Fördermethoden und Anstieg der Bohrungen. Da sich Amerika für das Fracking entschieden hat ist mit einem weiteren Anstieg der Reserven zu rechnen.
Häufig werden auch die technischen Möglichkeiten falsch eingeschätzt und es kommt zu Vorhersagen die dann nicht eintreffen. Ein Beispiel dafür ist die Gewinnung von elektrischer Energie durch Kernfusion. Liest man heute über die Zukunft der Fusionskraftwerke, so stehen diese kurz vor dem Durchbruch. In meiner Zeit als Student der Physik zu Beginn der 70 Jahre war dies genau dieselbe Botschaft die wir in unseren Vorlesungen zu hören bekamen.

Man kann also festhalten, die Kernfusion als unbegrenzte Energieressource war, ist und bleibt ein Zukunftsprojekt mit dem unbestreitbaren Charme die Energieprobleme dieser Welt final zu lösen.

3. Was gerne übersehen wird

Eines der ganz großen Themen ist heute der Klimawandel. Verschiedene Messdaten, z.B. aus Eisbohrkernen zeigen eine Korrelation von Temperatur und Kohlendioxidgehalt in der Atmosphäre. Plausibel scheint nun, daß der Mensch durch den vermehrten Kohlendioxidausstoß das Klima beeinflusst und deshalb durch eine Reduktion des Kohlendioxidausstoßes die globale Erwärmung gestoppt oder reduziert werden kann.
Das ist eine einfache für jedermann verständliche Aussage die einen förmlich dazu zwingt etwas gegen Kohlekraftwerke zu unternehmen und sich mit Claudia Roth von den Grünen zu solidarisieren, die am 27.4.13 sich wie folgt dazu äußerte:

„Aber die Politik dieser Regierung scheitert nicht nur draußen, in der Welt, sondern schon am eigenen Tellerrand –

- wenn Umweltminister Altmaier eine der größten Dreckschleudern in Europa einweiht –

- das Braunkohlekraftwerk Grevenbroich,

- mit jährlich 17 Millionen Tonnen CO2-Ausstoß,

- und das dann als herausragenden Beitrag zum Klimaschutz verkaufen will."

Man spürt hier förmlich wie böse, profitgeile Manger versuchen uns und unseren Kindern die Lebensgrundlage zu entziehen, es muß etwas geschehen!
So einfach liegen die Dinge aber nun leider nicht und ich möchte dies an einigen Beispielen aufzeigen:
Basis der Argumentation ist der ansteigende CO2 Gehalt in der Atmosphäre was außer Frage steht, oder doch nicht?
Man muß wissen, daß die Daten über den geschichtlichen CO2 Gehalt der Atmosphäre hauptsächlich aus Messungen von Lufteinschlüssen in Eisbohrkernen gewonnen werden. Ob die Ergebnisse die tatsächlichen damaligen Werte repräsentieren ist nicht unumstritten. So vertreten Jaworski et. al (2) die Meinung, daß es sich bei der Erhöhung der CO2 Anteile um natürliche Schwankungen handelt und in früheren Zeiten der CO2 Gehalt wesentlich höher war als laufend publiziert wird. Sie führen dazu ins Feld, daß bei früheren Veröffentlichungen der CO2 Befürworter Daten die höhere Werte hatten bei Berechnung des Anstieges nicht berücksichtigt wurden (2). Daß zudem alle publizierten Werte mehrfach gemittelt und bearbeitet wurden macht die Sache nicht einfacher. Travnicek et al. (3) zeigen, daß die Meßwerte sowohl Ortsabhängig stark schwanken, als auch über Jahreszeiten und im Tagesrythmus. Im worst case schwanken die Messwerte um bis zu 20 % um den Mittelwert.
Es gibt aber auch andere Stimmen, so weist Eric Monnin (4) in seiner Dissertation nach, daß bei geeigneten Präparationsverfahren der Fehler bei der CO2 Bestim-

mung aus Eiskernen durchaus gering gehalten werden kann. Wer nun die richtigen Werte präsentiert kann hier nicht entschieden werden. Es bleibt aber festzustellen, daß ein Problem existiert, das wissenschaftlich zu lösen ist und sich einer, wenn auch noch so demokratischen, Meinungsbildung entzieht.
Ein weiter Punkt betrifft die Umweltpolitik in der EU und ganz besonders in Deutschland. Immer wieder werden die deutschen Automobilhersteller attackiert und unsere Fahrzeuge als Spritfresser oder Dreckschleudern bezeichnet. Dabei wird so getan, als ob eine Reduktion des Verbrauches dem Klima zugute kommen würde. Dies ist eine weitere der „einfachen Wahrheiten“. Was sind aber nun die Fakten?
Die gesamten CO2 Emissionen betragen weltweit 8 * 10E11 t, davon sind 4 %, also 3,2 * 10E10 t antropogen, also vom Menschen verursacht. Deutschland hat davon einen Anteil von 8,76 *10E8 t, das entspricht etwa 0,1 % der gesamten CO2 Emissionen. Der Straßenverkehr ist daran mit 1,58 * 10E8 t also mit etwa 0,02 % beteiligt. Reduzieren wir den mittleren Verbrauch von heute ca. 160 g/km CO2 auf 95 g/km, würden wir unter Berücksichtigung des Gütertransportes den globalen CO2 Ausstoß gerade mal um ca. 50 ppm reduzieren. Natürlich macht es Sinn Ressourcen zu schonen, natürlich wollen wir den Kraftstoffverbrauch reduzieren aber daraus einen unverzichtbaren und signifikanten Beitrag zum Klimaschutz abzuleiten scheint mir wenig seriös.

Es ist unbestritten, daß sich das Globale Klima verändert, das zeigen fast alle zugänglichen Meßwerte, aber:

- Wir leben in einer Kaltzeit, während der letzten paar hundert Millionen Jahre war die globale Mitteltemperatur um fast 2 Grad höher als heute.

- Manchmal korreliert der Kohlendioxidgehalt mit der globalen Mitteltemperatur, manchmal nicht.

Ja, die Erde erwärmt sich, aber der Beweis steht aus, daß eine Reduktion des anthropogenen Kohlendioxidausstoßes dies ändern würde. Natürlich sollten wir den Kohlendioxidausstoß reduzieren, aber es ist genauso wichtig auch den Verbrauch von anderen Ressourcen zu reduzieren.
Dem Thema, ob es sinnvoll ist sich ausschließlich auf die Reduktion des antropogenen CO2 Ausstoß zu fokussieren will ich nun weiter nachgehen.

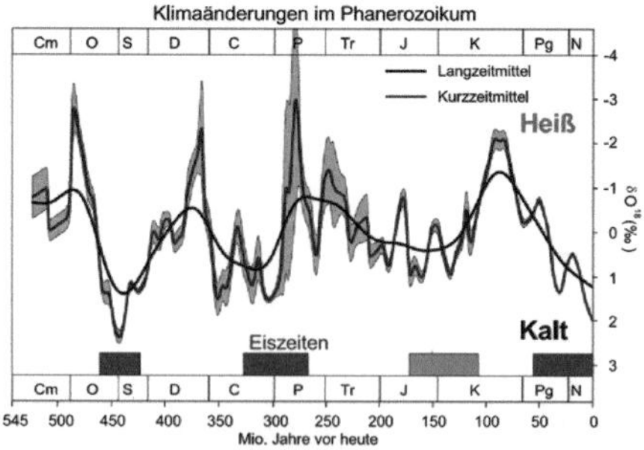

Bild 1: *Klimadaten (5)*

Was so alles grün ist

Smart Grid und erneuerbare Energien

Unser heutiges Stromnetz besteht aus großen Kraftwerken die gleichmäßig in Deutschland verteilt sind. Die elektrische Energie wird hauptsächlich durch rotierende Maschinen erzeugt, dabei ist die Leistung eines Turbinen/Generatorsatzes gegeben durch:

$$P = 2*\pi*M*n$$

wobei M das antreibende Moment und n die Drehzahl ist. Unser Netz wird mit einer Frequenz von 50 Hz betrieben, je nach Polzahl ergibt sich daraus die erforderliche Generatordrehzahl. Aus der obigen Gleichung ergibt sich nun, daß die Leistung im Netz von der Frequenz abhängt. Schnelle Änderungen der Last im Netz werden durch eine Änderung der Netzfrequenz beantwortet. Langsame durch die Anpassung der Antriebsmomente an den Generatorsätzen.
Bis zu einer Frequenzänderung von etwa 2 % bleibt das Netz stabil, darüber hinaus muß mit Momentenänderungen bis hin zum Erzeuger- oder Lastabwurfabwurf geregelt werden. Dies bedeutet in anderen Worten, daß das Netz innerhalb des genannten Frequenzbereiches sich bei schnellen Laständerungen selbst stabilisiert, da die rotierenden Massen der Generatorsätze als Speicher funktionieren. Größere Laständerungen können nur durch Momentenänderungen an den Generatorsätzen beeinflusst werden und das geht wesentlich langsamer. Kohle- und Kernkraftwerke können ihre Last um etwa 2 -4 % /min verändern, Gasturbinenkraftwerke etwa um 20 % der Nennleistung pro Minute.
Wenn wir nun die Grundlastkraftwerke immer mehr abschalten zugunsten der regenerativen Stromerzeuger, dann wird das Netz immer instabiler und die Gefahr von Abschaltungen steigt. Wir haben heute eine Regelreserve von im Mittel etwa 2 GW im Netz, die aber mit zunehmenden regenerativen Stromerzeugern abnimmt. Die Lösung des Problems heißt Smart Grid – so die landläufige Meinung. Was bedeutet dies? Die Idee dahinter ist, daß sowohl Erzeuger als auch Verbraucher in die Regelung eingebunden werden und dadurch die Netzstabilität aktiv hergestellt wird.

Hier gibt es aber eine Reihe von Risiken:

- Wir haben in Deutschland keine Möglichkeit eine Regelung des Netzes im erforderlichen Umfang außerhalb des Internets zu realisieren. Zwar wird immer wieder behauptet, daß durch geeignete Verschlüsselungsverfahren ein Zugriff von Unbefugten verhindert werden kann, doch scheint diese Argumentation immer brüchiger zu werden. Alleine über das Netz angeschlossene Stromzähler stellen schon ein erhebliches Risiko dar und werden nur zögerlich eingesetzt. Zudem arbeiten Heerscharen von Softwerkern in den Geheimdiensten an Cyberwaffen um im Krisenfall die gesamte Energieversorgung des Feindes zu stören, von irgendwelchen Hackern ganz abgesehen. Es wäre unverantwortlich ausschließlich auf ein solches System zu setzen.

- Ein weiteres Thema sind die Übertragungsverluste von Leitungen. Nach (6) beträgt der Verlust einer 380 kV Leitung ca 1,1*10E-4/km bei Nennlast, der einer HGÜ mit +/- 800 kV 3,5*10E-5/km (7). Während die thermischen Verluste der Leitungen bezogen auf die Umwelt einen eher vernachlässigbaren Einfluß haben, sieht dies bei den zusätzlichen Investitionskosten schon anders aus. Hier müssen die Verluste ja in Kraftwerksleistung vorgehalten werden. Bei 1000 km Leitungslänge und 5 GW Übertragungsleistung berechnen sich die Verluste zu:

380 kV Leitung	550 MW
HGÜ	175 MW

Die Verluste in den Umspannungsstationen sind dabei nicht eingerechnet.
Ein MW Windkraftwerk kostet heute etwa 1,5 mio € / MW, daraus lässt sich leicht der entsprechende Investitionsaufwand berechnen. Zudem zeigt die Abschätzung, wie sinnvoll es ist extrem lange Leitungen zu bauen, wie sie z.B. im Projekt DesertTec angedacht wurden.

- Ein Punkt bei der Smart Grid Diskussion ist die immer wieder diskutierte Speicherlösung mit Fahrzeugbatterien zur Netzstabilisierung. Die maximale Speicherfähigkeit der deutschen Pumpspeicherwerke beträgt heute etwa 38 GWh, die Zellenkosten für eine 1 kWh LiIonenbatterie in einem Auto betragen etwa 350 €, Targetpreis 2020 150 €. Umgerechnet wären dies dann 13,3 Mrd. € bzw. 5,7 Mrd. € Invest wollte man die Pumpspeicherkapazität durch Batterien ersetzen. Dies scheint allerdings nicht sinnvoll zu sein, man könnte die Batterien ja auch nur für Spitzenlast verwenden und nebenbei E-Autos betreiben. Der Ladungsdurchsatz einer LiIon Batterie beträgt etwa 2000 * Kapazität. Bei einem Zielpreis von 150 €/kWh würde allein der Speicherpreis für 1 kWh 15 Cent betragen, ohne Einrechnung der Verluste und der Überwachungselektronik. So etwas macht zumindest großflächig keinen Sinn und wäre ein neues Subventionsgrab.

Nun zu den erneuerbaren Energien

Erneuerbare Energien sind nicht wie uns der Name suggeriert erneuerbar und damit weitgehend kostenlos. Um die von der Sonne zur Verfügung gestellte Energie ernten zu können müssen technische Lösungen entwickelt werden und diese sind weder kostenlos, noch in jedem Falle umweltfreundlich. Zudem müssen noch Leistungsreserven vorgehalten werden, da die Verfügbarkeit der Ressourcen nicht oder nur begrenzt beeinflussbar ist. Für Deutschland ergibt sich für 2013 folgendes Bild (7):

Energieerzeuger	Installierte Leistung [GW]	Energiebeitrag [TWh]
Solar	35,65	29,7
Wind	32,5	47,2
Gas	27,2	39,4
Laufwasser	3,9	21
Steinkohle	24,9	110,6
Braunkohle	21,2	145,1
Kernenergie	12,1	92,1
Biomasse	7,15	48,3

Aus der Tabelle ist zu abzulesen, daß die installierten Leistungen und damit die Investitionen für Solar- und Windstromerzeugung zwar etwa 40 % betragen, der eingespeiste Energiebeitrag aber nur 13,5 % bezogen auf die Gesamtwerte 2013 in der Bundesrepublik. Grüner Strom bedarf also hoher Investitionskosten und ist folglich auf Länder begrenzt die sich dies leisten können.
Allein die ökonomische Betrachtung wird der Sache nicht gerecht, es gibt eine Reihe von Gründen für die erneuerbaren Energien die darüber hinaus geprüft werden sollten:

- Ressourcen

Nicht erst seit (1) wissen wir, daß die Ressourcen auf unserem Planeten begrenzt sind. Weniger dadurch, daß sie eines Tages nicht mehr vorhanden wären, das ist nicht zu erwarten. Vielmehr durch den immer höher werdenden Aufwand zu deren Gewinnung. In einigen Bereichen ist deshalb heute schon absehbar, daß der Umstieg auf regenerative Energien allein aus Kostengründen kommen muß. Dies gilt für Öl früher, für Kohle oder Uran später, aber es gilt. Ein Weg Ressourcen zu schonen ist dabei sicherlich ein verantwortungsvoller Umgang. Eine der unsinnigsten Verwendungen von fossilem Öl und Gas ist das beheizen von Häusern und es ist meist unserer Bequemlichkeit geschuldet. Nach Wikipedia wurden 2013 in Deutschland 55 % unseres Energiebedarfes durch Öl und Gas gedeckt, davon über die Hälfte zu Heizzwecken. Ökologisch sinnvolle (eben nicht nur dem k-Wert geschuldete Lösungen) Niederenergiehäuser in der Fläche würden eine höhere Ressourceneinsparung erbringen als die Abschaltung der Braunkohlekraftwerke.

- Emissionen

Die zunehmende Verstädterung bringt eine Reihe von erheblichen Problemen mit sich. Abgesehen von den Smogproblemen in den meisten „Megacities" besteht auch noch das Problem der Versorgung und Entsorgung des Abfallstoffe in solch riesigen Siedlungsräumen. Nach (8) sind z.B. in Neu Dehli fast flächendeckend multiresistente Bakterien im Abwasser und teilweise auch im Trinkwasser gefunden worden. Eine Gefahr, die unterschätzt wird und die ich für wesentlich bedrohlicher halte als die CO2 Emissionen aus Kohlekraftwerken.

- Klima

Wie bereits im vorigen Kapitel ausgeführt, ist die Klimaveränderung wahrscheinlich ein von uns kaum zu beeinflussender Vorgang. Es stellt sich hier natürlich die Frage, ob es wirklich sinnvoll ist alle Aufmerksamkeit der CO2 Reduktion zu schenken – was offensichtlich weltweit nicht funktioniert – und darüber alle anderen Risiken für die Menschheit mehr oder weniger zu verdrängen. Es gibt aber einen Weg die CO2 Emissionen mit existierenden Technologien zu reduzieren und damit sind wir beim Thema Kernkraft. Ein unglaublich emotionales Thema, erschreckt ja allein schon der Gedanke, daß der Abfall noch tausende, ja Millionen Jahre strahlt. Wie hoch ist aber nun tatsächlich das Risiko? Zum einen ist radioaktive Strahlung eine Gefährdung die einfach und präzise gemessen werden kann, schon in Größenordnungen weit unterhalb jeder Gefährdungsgrenze. Zum anderen ist die Technik von Kernkraftwerken weltweit unterschiedlich entwickelt, Deutschland gehörte zu den Ländern mit der am höchsten entwickelten Reaktortechnik. Und gerade wir schalten ab, fahren unsere Kerntechnologie herunter und glauben damit unser Risiko zu verringern. Leider macht radioaktive Verseuchung nicht an Grenzen halt, wie wir seit Tschernobil wissen. Es hilft auch nicht, daß wir glauben klüger zu sein, wenn China oder Finnland

Kernreaktoren bauen können wir nichts dagegen tun. Ein weiteres Argument ist die einfache Risikoabschätzung dessen was wir tun. Wer glaubt, Windkraftwerke und Mangos im Supermarkt sind ohne Risiko für menschliches Leben zu haben, der irrt. Die Frage ist einfach nur, wie hoch ist das Risiko und wählen wir unter all den Risiken klugerweise die geringsten.

- Verschwendung und Recycling

Wir verschwenden Energie in hohem Maße durch Transport von Menschen und Gütern, Klimatisierung unserer Habitate und nicht zuletzt in der immer mehr wachsenden Informationswelt. In Deutschland fließen mittlerweile zwischen 10 und 15 % der erzeugten Energie in die Informationstechnik, in USA und Japan noch wesentlich mehr. Googles Rechenzentrum in Portland ist der größte Einzelenergieverbraucher in den USA. Wie hoch ist der Nutzen und können wir uns das auf die Dauer leisten?
Ein weiteres Thema ist das Recycling, genau das was die Kernkraftgegner der Atomindustrie vorwerfen, nämlich die fehlenden Entsorgungsstrategien, genau das machen wir mit den erneuerbaren Energien.
Nach (9) fallen in den nächsten Jahren verstärkt große Mengen an Rotorblättern von Windkraftanlagen zur Entsorgung an. Die Mengen werden bis 2027 auf etwa 30 000 t Glasfaser- oder kohlefaserverstärktes Epoxidharz ansteigen. Von Seiten der Hersteller gibt es nahezu keine Konzepte der Entsorgung.
Bei den Speicherkonzepten wird die LiIonen Batterie immer als besonders grün gelobt. Pro Speicherkapazität von einer kWh benötigt man ca. 4 kg Elektrodenmaterial und etwa 1,7 l Elektrolyt. Im Elektrodenmaterial ist Ni, Mn, Co,.. enthalten, der Elektrolyt besteht aus Dimethylcarbonat, Ethykencarbonat, Propylencarbonat fein verteilt in den Elektrodenporen. Während die Elektrodenmaterialien durch geeignete Prozesse teilweise rückgewonnen werden können – was macht man mit der Ofenschlacke? – gibt es für die Elektrolytrückgewinnung keine umsetzbaren Verfahren.
Für die Entsorgung von Solarzellen, insbsonders der CIS Dünnschichtzellen gibt es Vorschläge die im Labormaßstab funktionieren.
Wie einfach ist es dagegen ein Kohlekraftwerk zu recyceln, Beton, Stahl, Kupfer und der Rest in den Ofen.

4. Zusammenfassung

Die Welt hat sich schon immer durch den Menschen verändert und sie wird es auch weiterhin tun. Das Problem dabei ist mittlerweile, daß unser Wissenstand so gewachsen ist, daß wir glauben mit hoher Genauigkeit Vorhersagen machen zu können. Aus diesen Vorhersagen will die Politik nun Maßnahmen ableiten, das ist ihre Aufgabe und dafür haben wir die Volksvertreter gewählt. Was aber nun, wenn sich unsere Vertreter und das Wahlvolk mit der Komplexität der Wirklichkeit schwer tun, wenn die Wissenschaft gar keine eindeutige Meinung hat? Wir müssen vereinfachen und glauben nur dadurch effizient handeln zu können, dies aber birgt Risiken. Möglicherweise haben sich Wissenschaftler geirrt oder es ist überhaupt noch nicht verstanden warum das Naturgeschehen so ist und nicht anders.

William von Ockham der Begründer wissenschaftlichen Vorgehens, sagt:
“Numquam ponenda est pluralitas sine necessitate”
Eine Mehrheit darf nie ohne Not zugrunde gelegt werden.

Aber es gilt auch und wäre im Sinne von William von Ockham, eine Vermutung ist nicht an sich schon deshalb richtig weil sie sich auf nur eine oder wenige Einflussfaktoren stützt.
Wie könnte denn nun unsere Zukunft aussehen?

Vier Thesen:

- Nicht ein Weg ist der richtige, sondern viele. Energieerzeugung durch einen Mix verschiedener Verfahren, Subventionen klar in Umfang und Laufzeit deckeln.

- Wenn wir die Zusammenhänge nicht verstehen werden wir auch die richtigen Entscheidungen nicht treffen können. Wir brauchen viel mehr Geld in die Grundlagenforschung (Mensch in der Umwelt, Zukunftsforschung, Globale Veränderungsprozesse).

- Wir müssen den Energieverbrauch von Information und Transport erheblich reduzieren.

- Wir müssen den Entwicklungs- und Schwellenländern behilflich sein Lebensbedingungen für ihre Bevölkerung zu schaffen, die vergleichbar mit unseren sind und dies so, daß wir auch danach noch eine globale Lebensgrundlage haben.

Lassen Sie mich mit einem Zitat schließen, das Albert Einstein zugeschrieben wird:

„Die reinste Form des Wahnsinns ist es, alles beim Alten zu lassen und gleichzeitig zu hoffen, dass sich etwas ändert."

Literatur

[1] The Limits to Growth, Dennis L. Meadows et al., 1972 (deutsch: Die Grenzen des Wachstums)

[2] Z. Jaworowski, T.V. Segalstad & V. Hisdal, Atmosheric CO2 and global warming: A citical review, Norsk Polarinstitutt, 1992, rev.2004

[3] W. Travnicek, S. Jacobi & M.Schmitt, 15 Jahre CO2 Messung in Hessen

[4] Eric Monnin, Die natürliche Variabilität der atmosphärischen Co2 Konzentration seit der letzten Eiszeit anhand von Messungen an antarktischen Eisbohrkernen, 2004

[5] Wikipedia

[6] B.R. Oswald, Verlust- und Energieabschätzung für das 380 kV Leitungsbauvorhaben Wahle – Mecklar, Hannover 2007

[7] V. Knack, Smart Grids, VDE München, 2012

[8] A.F. Widmer, medArt basel, 2013

[9] J. Woidasky, E. Seiler, Vortrag Hamburg T.R.E.N.D, 2013

3 A Dynamic Programming Based Simulation Tool for Optimizing Vehicle System Efficiency

Prof. **Melody Baglione**, Ph.D., The Cooper Union for the Advancement of Science and Art, New York, USA, **Mark J. Duty**, B.S.E.E.T, Chrysler LLC, Auburn Hills, Michigan, USA

Abstract

A dynamic programming (DP) algorithm, reverse tractive road load demand model, and Matlab®/Simulink® tool are developed to address the challenge of matching the powertrain hardware and control strategy for improved vehicle system efficiency. The DP-based methodology and reverse model optimizes transmission shift, torque converter lock-up, and multi-displacement cylinder deactivation control for consistent powertrain evaluation. The advantages of a DP-based simulation approach are demonstrated and key powertrain system integration concepts are revealed.

1. Motivation and Background

Simulation and virtual optimization is becoming more relied upon as vehicle system complexity increases and product development time decreases. Many simulations can provide efficiency and performance predictions but often require control calibrations, such as transmission shift maps, as inputs. A problem arises when analyzing future hypothetical hardware designs for which calibrations do not exist. A simulation approach based on dynamic programming (DP) is proposed for simulating powertrain hardware configurations with their optimal control strategies.

1.1 Vehicle System Modeling

Advanced vehicle models have become an essential tool to evaluate vehicle system performance during design. Forward-looking simulations include a driver model, often using a proportional-integral (PI) controller, which considers the desired vehicle speed to determine throttle and brake commands. The throttle command is translated into a fuel flow rate and engine torque, which is propagated through the drivetrain until the tractive force at the road and corresponding acceleration is calculated. In backward-looking models, the force required to achieve the acceleration is directly calculated from the desired speed trace. The force is then converted into the required torque and rotational speed that must be provided by the component directly upstream. This calculation approach is continued in the reverse direction of the road load force through the drivetrain until the necessary fuel energy is determined. Forward models have high fidelity but long run times and are often applied to performance simulation, hardware-in-the-loop, and to explore transient effects. Backward models are often preferred for fuel economy predictions, component sizing, and road load analyses due to their fast run times and simple integration routines. Further descriptions of both approaches and their advantages and disadvantages are discussed in [1]-[5]. A backward-looking approach was selected since a main goal is to quickly evaluate multiple design alternatives early in the design process.

Many virtual simulations, such as those described in [1]-[5], provide fuel economy predictions but often require control calibrations, such as transmission shift maps, as inputs. A problem arises when evaluating future design alternatives since calibrations do not exist for hypothetical hardware designs. Yet it is essential to modify the control strategy for different hardware designs to evaluate the full potential of the system under consideration. Consequently users must redesign the calibration, often subjectively, and manually input a hypothetical calibration for each design alternative into the simulation. This not only requires an extensive amount of time, once a simulation is run it is not known whether any potential benefit or loss in performance is related to the change in hardware or to the maturity of the calibration. A further complication arises since many methods to develop powertrain calibrations are subjective and do not take into account system interaction effects. A method to virtually determine an optimal control strategy offers great potential in advanced vehicle simulation.

1.2 Vehicle System Optimization

Model predictive control (MPC) originated in the late 1970's and employs a model to predict the output and calculate a control sequence by minimizing an objective function. Early vehicle design optimization work concentrated on optimizing engine control parameters [7]-[10]. Initial work by Wong, Clemens, and Porter extended the optimization problem beyond the engine to the driveline by incorporating powertrain matching techniques for improving fuel economy [11],[12]. DP was later applied to find the optimal gear shift sequence and the power split for a hybrid electrical truck [13]. DP combined with a forward-looking model was utilized to optimize shift maps for fuel economy based on constant throttle inputs [14]. This work further employs a model-based DP approach for simulating powertrain hardware configurations with their optimal control strategies.

2. Approach

2.1 Reverse Tractive Road Load Demand Model

A reverse tractive road load demand model simulates the powertrain in every possible state and a dynamic programming algorithm is applied to find an optimal control strategy. The reverse model is quasi-static, derivative based with inverted causality in that the road load force and corresponding wheel torque to achieve the acceleration is iteratively calculated from the vehicle speed trace. The road load force is modeled as a function of speed using the 3-term road load coefficients as described in Society of Automotive Engineers (SAE) standard J2263 [15]. The total required tractive force, F, is derived from the road load, vehicle inertia, and road grade forces:

$$F = A + Bv + Cv^2 + m\frac{dv}{dt} + mg\sin\theta \qquad (1)$$

where A, B, and C are the road load coefficients, v is the vehicle speed, m is the vehicle mass, g is the acceleration due to gravity, and θ is the angle the road makes with the horizontal.
The required torque and speed are then propagated through the powertrain system to the engine where the fuel flow is determined for all possible gear and torque converter states within the system hardware constraints. The vehicle subsystems modeled and direction of power flow are shown in Fig. 1. Driveline inefficiencies, parasitic losses, accessory loads, and component inertial losses as a function of speed and load are also accounted for during the power transfer calculations from component to component.

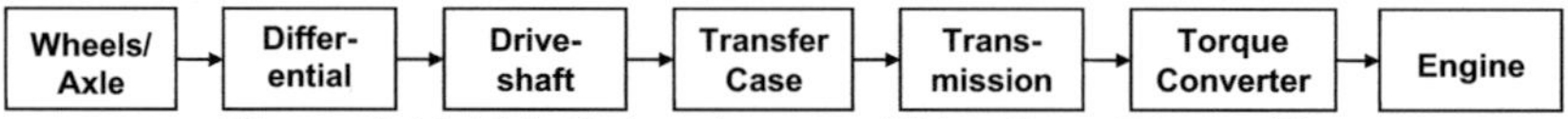

Figure 1: *Vehicle Subsystems and Direction of Power Flow*

2.2 Dynamic Programming Problem Formulation

The transmission shift and torque converter control strategies have a significant effect on the efficiency and overall fuel economy of the powertrain system. Once the propagated torque required to meet the road load demand reaches the transmission, the control strategy is treated as a multi-stage, multi-dimension decision process. In a stepped transmission, a number of gears may meet the road load demand at the prescribed transmission output speed; the challenge is determining which gear operates the engine at the lowest fuel flow with acceptable performance. Furthermore, automatic transmissions are typically equipped with torque converters and clutches that can be controlled to operate in an open, fully-locked (LU), or controlled partial slip state (PL), which also has a considerable impact on the overall system efficiency.

As the intention is to obtain a powertrain control strategy that minimizes the total fuel flow, $\dot{m}_f$, the objective function is:

$$J = \sum_{k=0}^{N-1} \dot{m}_f(k) \rightarrow \min \tag{2}$$

where N is the drive cycle duration and k is the time step. The objective function is subject to vehicle speed as well as engine and transmission speed and torque constraints as described in [2]-[6].

Gear shift scheduling is modeled as a discrete time dynamic system, where the shift is constrained by mechanical limitations of the stepped transmission. When the torque converter is in the open state, the impeller speed and torque are influenced by the design characteristics of the torque converter. The torque converter is characterized in the model by dynamometer data including its corresponding torque ratio, efficiency, and capacity or K-factor. K-factor is defined by:

$$K = \frac{N_e}{\sqrt{\tau_e}} \tag{3}$$

where K-factor is a function of speed ratio. The parameters used to describe the characteristics of a torque converter are depicted in Fig. 2.

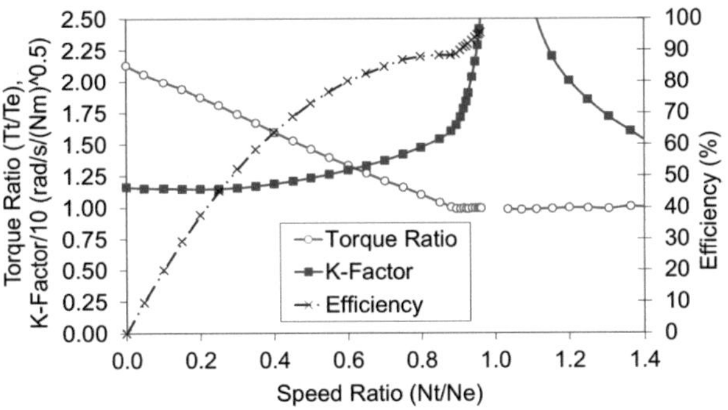

Figure 2: *Typical Torque Converter Characteristics*

Due to the model's reverse approach, the causality of torque converter K-factor relationship needs to be inverted and a similar K-factor relationship for the torque converter turbine is derived:

$$K_t = \frac{N_t}{\sqrt{\tau_t}} \tag{4}$$

where N_t is the turbine speed and τ_t is the turbine torque [16].

When the torque converter clutch is fully engaged, the torque converter's input and output shafts are locked, effectively eliminating any power loss yet losing any torque multiplication. In LU the engine torque is limited to the torque available at the corresponding turbine speed, which may be insufficient for acceleration performance. Also during LU, noise, vibration and harshness (NVH) due to torque fluctuations produced by the engine are transmitted through the driveline. For regions where torque fluctuation does not allow for full LU, electronically modulated control can be applied for PL to allow for some slippage which incurs some power loss but minimizes NVH. Clutch control constraints are properly integrated into the reverse model and DP algorithm as described in [3].

2.3 Dynamic Programming Algorithm

Discrete deterministic dynamic programming (DP) is applied as a model-based system design tool to find the control strategy that maximizes the powertrain efficiency over a desired drive cycle. DP is based on Bellman's Principle of Optimality and suggests that an optimal policy can be constructed in an iterative fashion by first solving the sub-problem at the last time step, N, then gradually extending the problem to include the last two time steps, and continuing in this fashion until the optimal policy for the entire problem is determined [17],[18]. The advantage of DP is that the optimal system state at a given time step is not viewed in isolation since control decisions will be ranked against the sum of the present costs and future costs, where the cost is the required fuel flow to meet the tractive road load demand. The cost function is additive in the sense that the cost incurred at time, k, accumulates over time. The total cost to be minimized is:

$$J = g_N(x_N) + \sum_{k=0}^{N-1} L_k(x_k, u_k) \tag{5}$$

where g_N is the cost at time step N, L_k is the instantaneous transition cost at time step k, and the system is modeled as a discrete time non-linear system of the form,

$$x_{k+1} = f_k(x_k, u_k), \quad k = 0, 1, \ldots, N-1 \tag{6}$$

where x_k defines the state of the system and u_k defines the control variables to be selected at time k.

The state and control variables are stored in matrices and the optimal global solution is determined by solving for the minimum cost recursively. The first step is to determine the minimum cost to go, J^*, from state, x_{N-1}, at time step, $N-1$,

$$J^*_{N-1}(x_{N-1}) = \min_{u(N-1)}[g_N(x_N) + L(x_{N-1}, u_{N-1})] \tag{7}$$

given the instantaneous transition cost at $N-1$ for each decision, u, and continuing backwards in time for $0 < k < N-1$,

$$J^*_k(x_k) = \min_{u(k)}[J^*(x_{k+1}) + L(x_k, u_k)] \tag{8}$$

until the first step is reached and the optimal control sequence and minimum accumulated cost for the entire cycle duration is determined.

The DP simulation fuel economy results are compared to experimental chassis dynamometer measurements in [2]-[5]; these comparisons reveal the DP simulation can reliably predict potential fuel economy benefits.

3. Application

3.1 Powertrain System Integration

Fuel economy is highly sensitive to vehicle attributes. Minimizing weight, for instance, not only has a direct benefit on fuel economy but also yields a further benefit by reducing the demand requirements of the engine. One advantage of a DP-based simulation approach over other vehicle models is that as vehicle parameters are changed, the DP algorithm automatically takes corresponding system level effects into account. For example, as vehicle weight is reduced, the transmission control strategy is automatically adjusted to upshift earlier so the engine can operate at a lower engine speed with less fuel energy demand.

Various vehicle and powertrain configurations were evaluated over the U.S. Federal Test Procedure (FTP) Urban Dynamometer Driving Schedule (UDDS) using the DP simulation tool. Since modifying the powertrain design to improve fuel economy can come at a detriment to acceleration performance, a forwarding-looking wide open throttle simulation was integrated into the analysis as described in [3]. For purposes of this study, brake specific fuel consumption data was scaled to determine the torque and fuel flow characteristics of different displacement engines. Only vehicle configurations with comparable ¼-mile (~400m), 0-to-60 mph (~0-100 km/hr), and 0-to-30 mph (~0-50 km/hr) times were considered. The DP optimized fuel economy results for vehicles with similar acceleration performance are compared in Fig. 3 to illustrate the sensitivity of changes in vehicle weight.

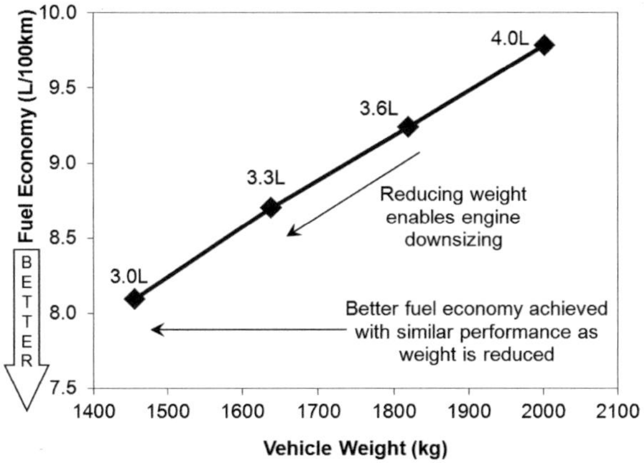

Figure 3: *Effect of Vehicle Weight and Downsizing*

3.2 Vehicle System Efficiency Analyses

Only a fraction of the total fuel energy goes into propelling a vehicle down the road. A vehicle system efficiency metric is proposed in addition to fuel economy in terms of liters per 100km (or miles per gallon). Vehicle system efficiency is automatically output by the DP-based simulation tool and is defined as the propulsion energy over the total fuel energy, as shown in Eqn. 9:

$$\text{VehicleSystemEfficiency}, VSE(\%) = \frac{\int F_P v dt}{\int \dot{m}_f dt \cdot Q_{NHV}} \cdot 100 \tag{9}$$

The propulsion energy for a vehicle is obtained by integrating the propulsion power, which is determined by multiplying the propulsion force, F_P, by the vehicle's velocity. The propulsion force is the net positive sum of the road load, inertial, and grade forces as defined in Eqn. 1. The fuel energy is obtained by integrating the fuel flow and multiplying by the net heating value of fuel, Q_{NHV}.

To minimize fuel usage, vehicle subsystems must be made to operate as efficiently as possible. However, it is equally important that each component interact with the system so as to maximize the efficiency as a whole. For advanced vehicle designs, it is crucial that the engine displacement, torque converter, transmission and final drive ratio match the vehicle system. Once target vehicle attributes are established, the DP optimization tool can be used to match the powertrain to specified vehicle attributes.

Three transmissions with different ratio spreads were analyzed with four engine displacements for a target sedan over the UDDS. Descriptions of the transmissions considered can be found in [19],[20]. To reflect the transmission inefficiencies in the analysis, theoretical gear efficiencies were assumed based on the gear ratio and clutch configuration. Spin losses (i.e., losses in open running clutches) were assumed based on the number of discs, clutch diameter, gear, geometry, and engine speed. The results in Fig. 4 indicate that more transmission gears and a wider ratio spread is desirable since it enables downsizing engine displacement to achieve higher vehicle system efficiency with constant or better performance. Note, for the vehicle under consideration a 2% increase in vehicle system efficiency corresponds to a 10% improvement in fuel economy (or a reduction of 0.92 L/100km).

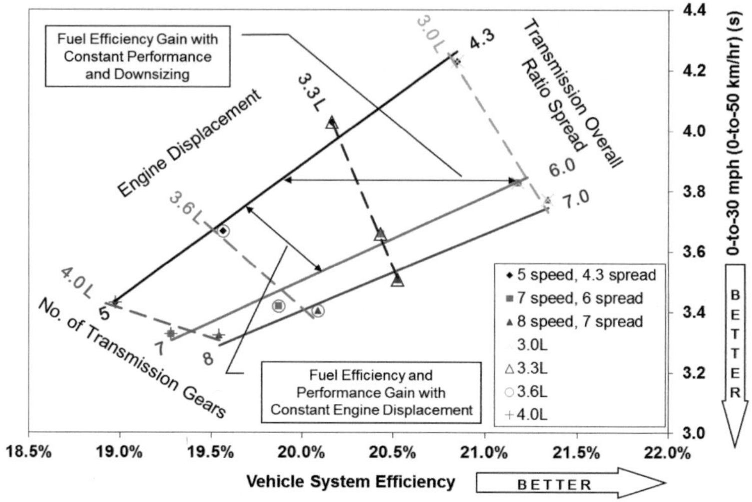

Figure 4: *Effect of Engine Displacement, Transmission Gears and Ratio Spread on Performance and Vehicle System Efficiency*

Final drive ratios were selected for each configuration in Fig. 4 by sweeping a number of ratios and creating a performance and fuel economy "hook" similar to Fig. 5. A near optimum final drive ratio was selected such that lower numerical ratios yield no appreciable gain in fuel efficiency for a loss in performance and higher ratios produce no appreciable performance gain for a loss in fuel efficiency. To simplify the study similar axle efficiencies were assumed, although in practice slight increases in efficiency may result with lower axle ratios.

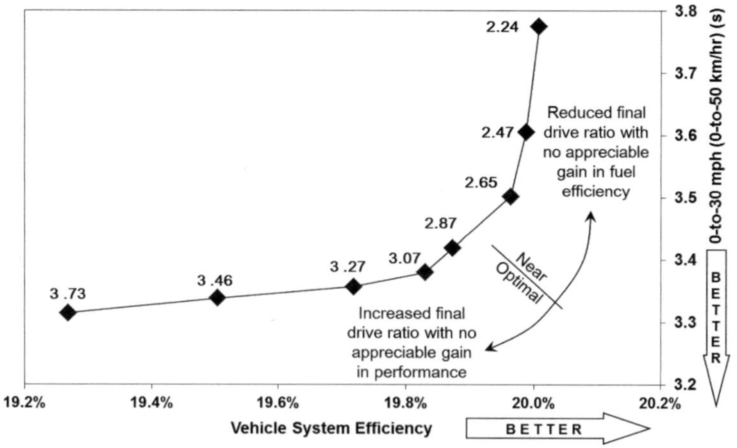

Figure 5: *Near Optimal Final Drive Ratio Selection*

One primary advantage of a DP simulation approach is the significant time savings to evaluate multiple powertrain design alternatives. The analysis in Figs. 4 and 5 required simulations for a total of 96 powertrain configurations (4 engines × 3 transmissions × 8 final drive ratios). Matlab® code automatically changes the input parameters for the powertrain configurations of interest and loops the simulations thereby expediting the total simulation time. The simulation time for the 96 configurations is approximately 40 times faster than real time. In addition to computational time savings, additional time savings result by using the DP algorithm since new powertrain control inputs (e.g., new shift maps for each configuration) do not need to be developed offline when changing the powertrain configuration. Between the computation time, control development and input time, the DP-based simulation tool reduces the analysis time from a number of days to around an hour.

3.3 Multi-Displacement System Effects

Multi-displacement system (MDS) is a proven technology for improving fuel economy without sacrificing performance [21]. MDS functions by deactivating the intake and exhaust valves and shutting off fuel for some of cylinders under part load. MDS was incorporated into the DP simulation model to better facilitate MDS integration with the appropriate driveline configuration and calibration strategy [22]. Fig. 6 demonstrates the UDDS fuel efficiency and acceleration performance effects of adding MDS engine technology to various engine and transmission combinations.

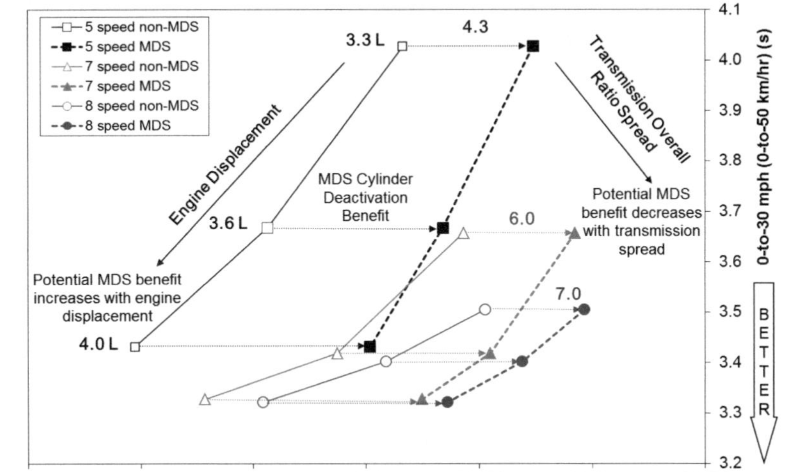

Figure 6: *MDS Effect on Vehicle System Efficiency*

MDS technology achieves higher fuel efficiency while maintaining the same performance. Given that MDS reduces pumping work by deactivating half of the engine cylinders under part load, the percent fuel efficiency benefit lessens as the displacement is reduced. Smaller displacement engines have less torque available with some of their cylinders deactivated and maintain less cylinder deactivation time. Increasing the number of transmission gears and ratio spread enables the engine to operate in a more efficient region such that the percent benefit due to MDS is less significant. A benefit of the DP approach is its ability to optimize MDS operation while taking gear and clutch control interaction effects into account; for instance, when higher torque is demanded the DP algorithm determines whether the control system should exit MDS operating mode or downshift and maintain MDS operation.

This analysis suggests that achieving higher fuel efficiency requires the appropriate combination of new technology as adding many features that attempt to reduce the same losses (e.g., pumping work) will not result in additive fuel efficiency benefits. Likewise, when estimating vehicle system efficiency, the individual measured fuel efficiency benefit from higher speed transmissions, MDS, or other technologies, cannot be summed together when combined. This further demonstrates the advantage of the DP simulation approach since the interaction effects of individual technologies are taken into account to determine the combined effect.

4. Conclusion

A DP simulation approach facilitates virtual optimization of powertrain control for the design of fuel efficient vehicle systems. By catering the shift, lock-up, and MDS control strategies to given powertrain configurations and drive cycles, DP facilitates rapid and consistent evaluation of design alternatives. The DP algorithm optimizes control taking system interaction effects into account. As the demand for more fuel efficient vehicles increases, DP can help to quickly determine optimal combinations of powertrain technology in conjunction with optimized control strategies.

References

[1] Markel, T., Brooker A., Hendricks, T., Johnson, V., Kelly, K., Kramer, B., O'Keefe, M., Sprik, S., Wipke, K, "ADVISOR: A Systems Analysis Tool for Advanced Vehicle Modeling," *J. of Power Sources*. 110: 255–266, 2002.

[2] Baglione, M., Duty, M.J., Ni, J., Assanis, D., "Reverse Dynamic Optimization Methodology for Maximizing Powertrain System Efficiency," *Proc. of the Fifth IFAC Symposium on Advances in Automotive Control*, 5(1), Aptos, CA, Aug. 20-22, 2007.

[3] Baglione, M., "Development of System Analysis Methodologies and Tools for Optimizing Vehicle System Efficiency," Ph.D. Dissertation,Univ. of Michigan, 2007.

[4] Baglione, M., Duty, M.J., "Development of Reverse Dynamic Optimization Methodology for Optimal Powertrain Integration and Control Design". *Proc. of the ASME International Design Engineering Technical Conferences & Computers and Information in Engineering Conference*, Brooklyn, NY, Aug. 3-6, 2008.

[5] Duty, M.J., Baglione (formerly Papke), M., "Methods and Systems for Powertrain Optimization and Improved Fuel Economy," U.S. Patent 8,050,856, Nov. 1, 2011.

[6] Baglione, M., Duty, M.J., "Development of a Powertrain Matching Analysis Tool," *SAE Technical Paper 2010-01-0490*.

[7] Auiler, J.E., Zbrozek, J.D., Blumberg, P.N., "Optimization of Automotive Engine Calibration for Better Fuel Economy – Methods and Applications," *SAE Technical Paper 770076*, 1977.

[8] Kolmanovsky, I., Sivashankar, S., Sun, J., "Optimal control-based powertrain feasibility assessment: A software implementation perspective," *Proc. of the 2005 American Control Conference*, 7: 4452- 4457, 2005.

[9] Kolmanovsky, I., Van Nieuwstadt, M., Sun, J., "Optimization of complex powertrain systems for fuel economy and emissions," *Proc. of the 1999 IEEE International Conference on Control Applications*, 1: 833-839, 1999.

[10] Kang, J.M., Kolmanovsky, I., Grizzle, J.W. "Approximate dynamic programming solutions for lean burn engine aftertreatment," *J. of Dynamic Systems, Measurement, and Control*. 123: 153–160, 2001.

[11] Wong, L.T., Clemens. W.J., "Powertrain Matching for Better Fuel Economy," *SAE Technical Paper 790045*, 1979.

[12] Porter, F.C.,"Design for Fuel Economy – The New GM Front Drive Cars," *SAE Technical Paper 790721*, 1979.

[13] Lin C.C., Peng, H., Grizzle, J.W., Kang, J.M., "Power management strategy for a parallel hybrid electric truck," *IEEE Transactions on Control Systems Technology*, 11:839- 849, 2003.

[14] Kim, D., "Math-Model Based Gear-Shift Control Strategy for Advanced Vehicle Powertrain Systems," Ph.D. Dissertation, Univ. of Michigan, 2006.

[15] SAE Standard, "Road Load Measurement using Onboard Anemometry and Coastdown Techniques," J2263, 2008.

[16] Blumberg, P.N., "Powertrain Simulation: A Tool for the Design and Evaluation of Engine Control Strategies in Vehicles," *SAE Technical Paper 760158*, 1976.

[17] Bellman, R., *Dynamic Programming*, 6th ed., Princeton Univ. Press, New Jersey, 1972.

[18] Denardo, E.V., *Dynamic Programming: Models and Applications*, Prentice-Hall, New Jersey, 1982.

[19] Greiner, J., Doerr, C., Nauerz, H., Graeve, M., "The New 7G-TRONIC of Mercedes-Benz: Innovative Transmission Technology for Better Driving Performance, Comfort, and Fuel Economy," *SAE Technical Paper* 2004-01-0649.

[20] Wagner, G., H. Naunheimer, H. Scherer, A. Dick, "Achtgang-Automatikgetriebe zur Reduzierung des Kraftstoffverbrauchs," Automobiltechnische Zeitschrift, 512-519, June 2007.

[21] Falkowski, A., McElwee, M. Bonne, M., 2004. "Design and Development of the DaimlerChrysler 5.7L HEMI® Engine Multi-Displacement Cylinder Deactivation System," *SAE Technical Paper 2004-01-2106*.

[22] Baglione, M., Duty, M.J., "Reverse Dynamic Optimization of Variable Displacement Engine Operation and System Integration," *Proc. of the ASME Dynamic Systems and Control Conference*, Ann Arbor, MI, Oct. 20-22, 2008.

Acknowledgements
The author acknowledges former Chrysler colleagues Greg Pannone and Bob Lee for supporting this work and Prof. Jun Ni and Prof. Dennis Assanis for their guidance.

4 Cost-optimal Operational Strategies for Hybrid Vehicles

Dipl.-Ing. **Johannes Fuchs,** Dr. tech. **Bernhard Schweighofer**, Dr. techn. **Hannes Wegleiter,** Institute of Electrical Measurement and Measurement Signal Processing, Graz University of Technology, Graz, Austria

Abstract

This paper presents a new approach for cost-optimal operational strategies for hybrid electric vehicles. Therefore we calculate an operating strategy which minimizes the running costs during the whole vehicle life span. These costs are mainly described by the sum of fuel costs and costs due to the exchange of components. In our approach we focus on the replacement of the battery and mechanical brakes since the wearing of these components can be directly influenced by the operational strategy. We use “deterministic dynamic programming” (DDP) to calculate the minimal running costs. Additionally, this algorithm delivers the global optimal operational strategy for a given hybrid vehicle configuration. Finally, we demonstrate our method on a hybrid bus model and simulate different scenarios in form of a small parameter study.

1. Introduction

Hybrid vehicles provide a high potential to improve the efficiency and reduce emissions. In contrast to conventional vehicles, there are at least two power sources to cover the power demand of different load cases. For the power distribution between the power sources an operational strategy is needed. To optimize the operational strategy of a hybrid vehicle, several conditions have to be taken into account. These are mainly the size of components (e.g. energy storage) and the efficiency of energy converters. In many applications the minimization of fuel consumption is the only goal of an operational strategy. In contrast to this standard approach in this paper we focus on a strategy that minimizes the running costs in terms of money for hybrid vehicles. For this purpose we consider fuel costs and costs for exchanging the battery and mechanical brakes. Our method could be interesting for public transport or freight traffic since in these applications the running costs are more important than the fuel consumption. We use “deterministic dynamic programming” (DDP) for calculating the optimal control strategy under different target cost functions. One big advantage of the DDP-algorithm lies in the fact, that it calculates the global optimal solution. Hence, it is possible to analyse different vehicle configurations in a “fair” way since possible local minima can’t influence the comparison. In addition, the DDP-algorithm delivers the global optimal operational strategy for a given parameter set. Although the calculated control strategy is acausal, it can be used as a benchmark for heuristic approaches. In the following chapters we explain our approach in detail and further demonstrate the method on basis of a small example.

2. The Dynamic Programming Algorithm

The dynamic programming algorithm was introduced by Bellman [1] and can be used in many scientific areas. The algorithm calculates the global optimum of a mathematical problem in an acceptable computing time. The reduction of calculation effort is achieved by dividing a main problem into smaller subproblems and the re-use of calculated solutions. Thus, optimization problems with exponential calculation effort can be reduced into polynomial complexity.
The following subsections present our approach for calculating the minimal running costs of hybrid vehicles and additionally show some practical hints in the field of dynamic programming.

2.1 Basic Dynamic Program for Hybrid Vehicles

In general a dynamic program consists of a target cost function, state variables and control variables. It has been shown in previous work [2] that the DDP-algorithm can be applied in the field of optimizing hybrid electric vehicles. The target cost function contains costs, which are often defined as the total fuel consumption. In this paper we focus on the minimization of running costs in terms of money. These costs are approximated by a sum of fuel costs and costs for exchanging the battery and brakes due to wearing effects. To calculate these costs the state and control variables have to be appropriately defined. We choose the state of charge (SoC) of the battery as state x and the battery power as control variable u.

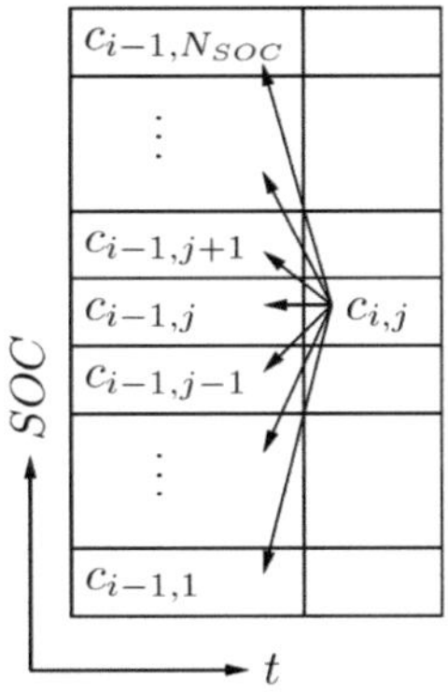

Figure 1: *Dynamic programming calculation step for hybrid vehicles [3]*

Figure 1 depicts the principle of a calculation step. The costs $c_{i,j}$ at a time-index i and SoC-index j are the minimum value of the sum of costs at prior states $c_{i-1,x}$ (time-index = i-1, SoC state x = 1 to N_{SOC}) and costs for reaching the SoC-index j at time i starting from the corresponding state x.

2.2 Consideration of nonlinear Costs

For each calculation step the transition costs from state $x_{i-1,j1}$ to $x_{i,j2}$ have to be determined. These costs depend on the control variable $u_{i,j}$ and on the prior state $x_{i-1,j1}$. In our approach we use a control variable u_i which is a function of the state changes Δx ($x_{i,j2}$-$x_{i-1,j1}$). In addition, we have to investigate the defined cost dependency in more detail. In general, costs can be described as a linear or nonlinear function. Linear costs, which are shown in figure 2, only depend on Δx. That means that for a given Δx the same costs ΔJ are caused and the absolute value of the previous state variable $x_{i-1,j1}$ doesn't influence the cost calculation. In the nonlinear case another approach has to be chosen to avoid significant errors. Figure 3 shows the influence of equal state changes (Δx) on the calculated costs. It can be seen that the transition costs do not only depend on Δx but also on the absolute value of the previous state $x_{i-1,j}$.

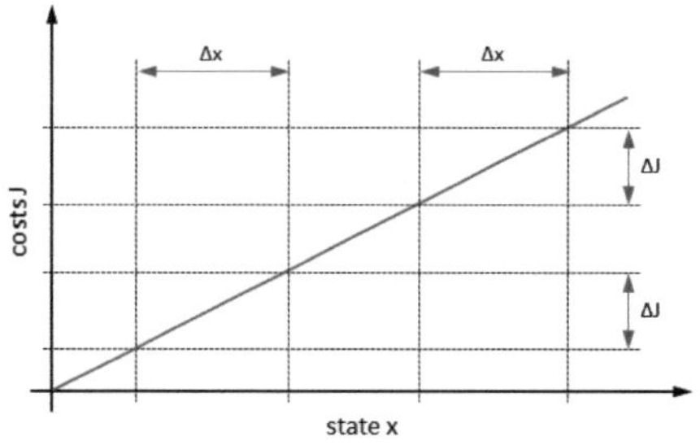

Figure 2: *Linear costs*

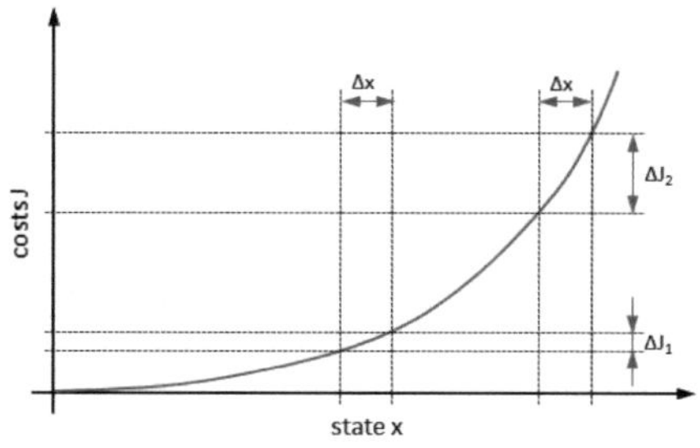

Figure 3: *Nonlinear costs*

2.3 Practical Adaptations

The grid discretization strongly influences the quality of the solution and has to be chosen carefully. A curse discretization leads to a fast computation time but also to a lower accuracy. By using a fine discretization the advantage of the polynomial complexity of the DDP-algorithm gets lost and problems become infeasible. In literature this effect is often described as the "curse of dimensionality" [4,5].

One practical method for calculating an accurate solution with a fine discretization is the use of an adaptive grid. This method is proposed amongst others by [5, 6] and can be applied for reducing the calculation effort. The idea of this technique is that the first solution of the dynamic program is calculated with a coarse discretization. This solution provides a reference trajectory which is used for the adaptation of the grid. For the next iteration steps a smaller grid with a finer discretization is built around the previous calculated trajectory. If the changes of the trajectory between two iterations lie below a threshold the procedure can be stopped. Instead of calculating the whole state space with a fine discretization the proposed method reduces the discretization only in a part of the grid and leads to a lower total computation time in spite of the iteration process. The reduction of the band around the reference trajectory has to be investigated before starting the iterations. If the band is too small, the search space for the DDP-algorithm is restricted and the calculated solution can diverge from the actual global optimum.

Figure 4 shows solutions of different iteration steps. It can be seen that the first iteration delivers an approximate estimate of the optimal trajectory. With increasing iterations and grid adaptations the solution converges towards the global optimum.

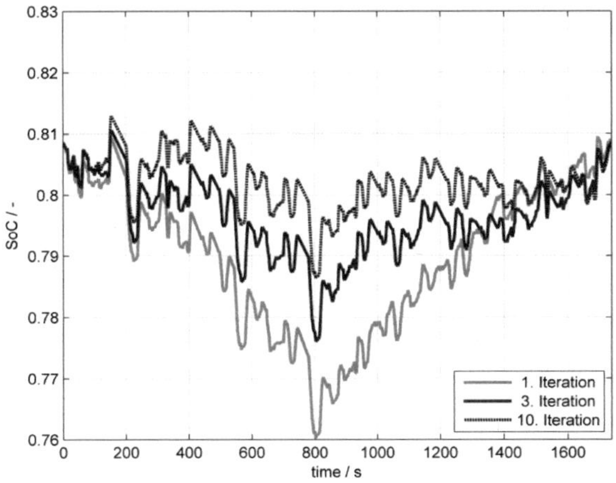

Figure 4: *Convergence of SoC-trajectories*

3. Hybrid Vehicle Configuration

In order to demonstrate our approach a model of the Volvo 7700 hybrid bus in combination with a serial hybrid topology is used. In figure 5 the general topology of a serial hybrid electric vehicle is shown.

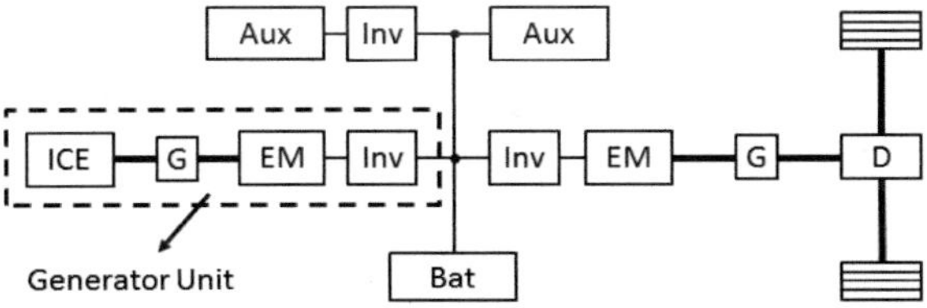

Figure 5: *Topology of a serial hybrid vehicle [7]*

ICE	Internal Combustion Engine	Inv	Inverter
G	Gearbox	Aux	Auxiliary devices
D	Differential	▬	Electrical connection
EM	Electrical Machine	—	Mechanical connection
Bat	Battery		

The basic parameters of the modelled bus are shown in table 1. The maximum powers of the generator unit and the electrical machine of the drivetrain are defined with 150 kW and 320 kW, respectively.

Table 1: *Parameters of the Volvo 7700 Bus [7]*

mass	18.900 kg
drag coefficient	0,6
reference area	8,1 m²
roll friction coefficient	0,0075

In figure 6 the linear behaviour between the cumulative brake energy and brake wear under different temperatures is shown [8]. We use this brake wear model in our approach to calculate the braking costs. Since it shows a linear behaviour, no additional dimension in the DDP-algorithm is needed and so a higher computation complexity can be avoided.

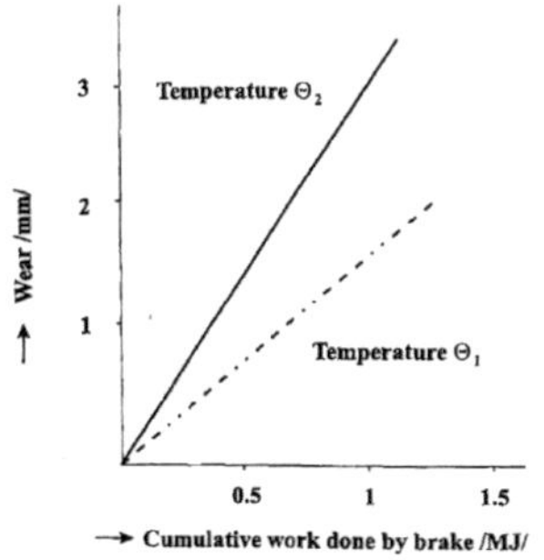

Figure 6: *Applied brake model [8]*

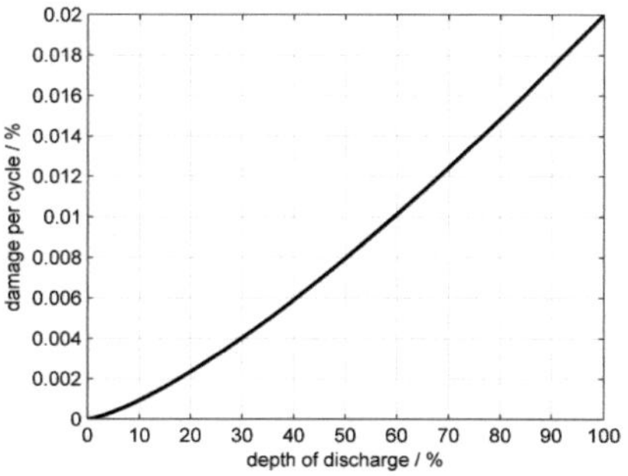

Figure 7: *Battery damage depending on the depth of discharge [7]*

The battery damage model, which is shown in figure 7, shows a nonlinear relation between the damage per cycle and the depth of discharge [7]. As described in section 2.2 this behaviour has to be especially considered in the DDP-algorithm. To maintain the global optimal solution it is necessary to use an additional storage in the state grid. With the help of this storage the total values of occurring charge and discharge swings are stored and so the nonlinear influence of relative changes in the SoC can be considered. The DDP-algorithm delivers battery costs which only describe the battery damage due to cycle effects. In order to use a more realistic battery damage model we apply these cycle effects in combination with calendaric battery damage effects. Figure 8 shows these two relations which affect the life span of a battery. Since we don't know the impact on the battery damage of both mentioned effects in advance, we use an iterative approach to obtain the solution for the total battery costs.

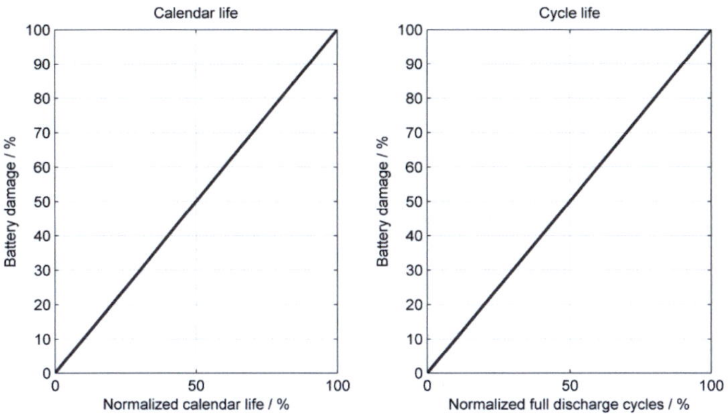

Figure 8: *Considered battery damage effects*

4. Simulation Results

In this section the results of a small parameter study are shown. As described in section 3 we use a model of a serial hybrid bus for demonstrating the approach.

4.1 Driving Cycle

We choose the "Braunschweig City Driving Cycle" as driving cycle since this profile represents a typical bus application. With the help of the vehicle parameters it is possible to convert this profile into time-depending power demands. In the propulsion case positive power demands have to be covered by the generator unit or the battery. Negative powers which represent braking are provided by the mechanical brake or the battery. The optimal power split, which is the main task of an operational strategy, is calculated with DDP for different parameter sets.

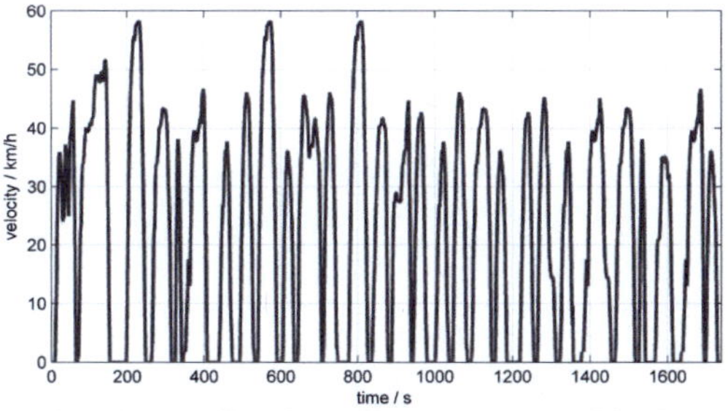

Figure 9: *Velocity profile of the "Braunschweig City Driving Cycle"*

4.2 Parameter variations and simulation results

To investigate different influences on the simulation results, a small parameter study was performed. The variable parameters were fuel costs, battery costs, battery energy contents, brake exchange costs and operating hours during a life cycle.
For comparing all scenarios we define the total life cycle of the bus as 10 years. In addition we scaled down all calculated costs to one driving cycle to avoid the influence of different operating hours per year. We also considered auxiliary devices (e.g. air conditioning) in form of an additional power demand. In our approach we assume a constant power of 15 kW, but a time-varying power demand could be used as well. We consider fuel costs, battery costs and braking costs for calculating the minimum running costs of the hybrid electric vehicle. Therefore, it is necessary to define some typical weighting factors in the cost functional. The calculated fuel consumption can be easily converted into costs. The steps for determining the battery costs are described in the previous section 3. Additionally, we defined the end-of-life time of the battery with 10 years to describe the calendaric aging effects. The maximum number of full discharge cycles is assumed with 3000 at an ambient temperature of 25°C. For the braking costs we suppose that the mechanical brakes could endure three years at high loads (5000 hours/year). The total costs for exchanging the brakes are defined with 5000 €. With this assumptions and the used brake model [8] it is possible to convert different braking energies directly into costs.
In figure 10 and 11 the calculated power split for a section of the driving cycle is shown. The parameter sets contain fuel costs of 1 €/l, a battery energy content of 20 kWh and 5000 operating hours/year but different battery costs (500 and 2000 €/kWh).

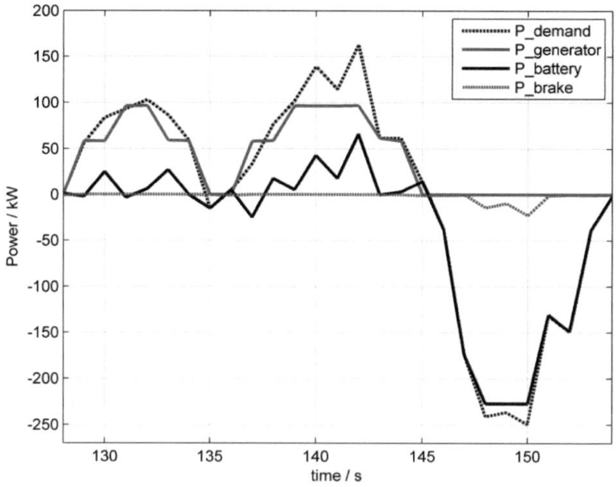

Figure 10: *Power split for a section of the driving cycle (fuel costs = 1 €/l, battery energy content = 20 kWh, 5000 operating hours/year, battery costs = 500 €/kWh)*

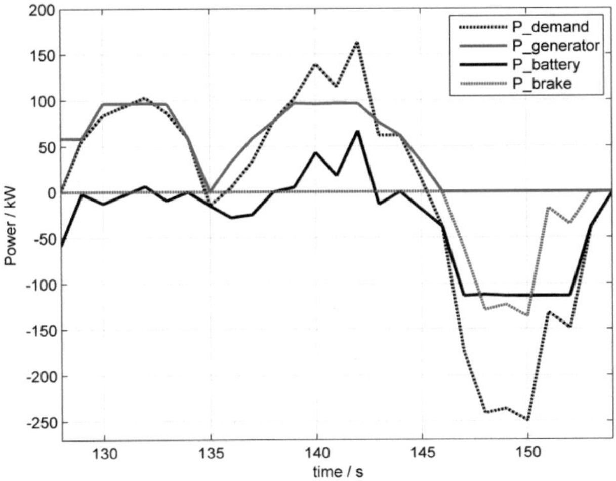

Figure 11: *Power split for a section of the driving cycle (fuel costs = 1 €/l, battery energy content = 20 kWh, 5000 operating hours/year, battery costs = 2000 €/kWh)*

Figure 10 and 11 show that the sum of generator power and battery power covers the power demand in the propulsion case. In the braking case the negative power is provided by the sum of the brake and battery power. If we compare both figures we can see that higher battery costs cause a preferred usage of the mechanical brake.

In table 2 the minimal costs for one driving cycle and the corresponding contribution of each component are shown for different parameter sets. In the first row the basic parameters are listed. In the further rows only one parameter was varied to allow a more meaningful comparison of the results. It can be seen that the braking costs are very small compared to the total costs. In these scenarios the costs due to the brake wearing per driving cycle can be neglected. The main impact on the total costs is caused by the fuel costs per litre and battery costs per kWh. Both parameters have a huge influence on the split between fuel and battery costs.

Table 2: *Simulation results for different parameter sets*
First row: reference case, remaining rows: parameter study (changed parameters in bold)

Parameter set				Costs per driving cycle			
Fuel costs	Battery costs	Battery energy content	Operating hours	Total	Fuel	Battery	Brake
€ / litre	€ / kWh	kWh	h/year	€	€	€	ct
1	500	20	5000	7,553	6,85	0,67	3,3
3	500	20	5000	21,213	20,49	0,69	3,3
1	**2000**	20	5000	9,562	7,03	2,50	3,2
1	500	**40**	5000	7,015	6,32	0,69	0,5
1	500	20	**2000**	7,683	6,85	0,80	3,3
1	0	0	**5000**	11,241	11,08	0	16,1
				Total life cycle costs			
Fuel costs	Battery costs	Battery energy content	Operating hours	Total	Fuel	Battery	Brake
€ / litre	€ / kWh	kWh	h/year	t€	t€	t€	€
1	500	20	5000	781,7	709,0	69,3	3416
3	500	20	5000	2195,5	2120,7	71,4	3416
1	**2000**	20	5000	989,7	727,6	258,8	3312
1	500	**40**	5000	726,0	654,1	71,4	517
1	500	20	**2000**	318,1	283,6	33,1	1366
1	**0**	**0**	5000	1163,5	1146,8	0	16667

To compare the total life cycle costs of different parameter sets the calculated costs per driving cycle were appropriately scaled up. The results show that braking power is preferably delivered by the generator (recuperation case) and so the total mechanical brake costs tend to be low. For the total battery costs a more realistic approach should be considered since only the exchange of an integer number of batteries can occur during a life cycle. In case of the scenario in the first row seven batteries have to be used. This would actually lead to total battery costs of 70 t€. Additionally, a comparison between hybrid and conventional vehicle topologies is shown.

5. Conclusion

In this paper we demonstrated a new approach for calculating cost-optimal operational strategies for hybrid vehicles. For this task we used DDP since this algorithm allows finding the global optimum of a problem in an acceptable computation time. In our cost function we consider costs in terms of money which consist of a sum of fuel costs, battery costs and brake costs. For the demonstration of our method we used a model of a serial hybrid bus and performed a small simulation study to investigate the influence of different parameter sets. With the help of the proposed method it is possible to calculate a benchmark for cost-optimal operational strategies under different conditions.

Literature

[1] R. E. Bellman. Dynamic programming, Princeton - N.J., Princeton University Press, 1957.

[2] O. Sundström. Optimal Control and Design of Hybrid-electric Vehicles, Dissertation, ETH Zürich, 2009.

[3] H. Wegleiter, B. Schweighofer, G. Holler, R. Brunnader. Fast Quasi Optimal Control of Hybrid Electric Vehicles considering Limiting Conditions, European Electric Vehicle Congress EEVC, 2011.

[4] D. Bertsekas. Dynamic Programming and Optimal Control, Athena Scientific, 2005.

[5] R. Luus. Iterative Dynamic Programming, Chapman & Hall/CRC, 2000.

[6] H. G. Wahl, F. Gauterin. An Iterative Dynamic Programming Approach for the Global Optimal Control of Hybrid Electric Vehicles under Real-time Constraints, IEEE Intelligent Vehicles Symposium (IV), 2013.

[7] J. Fuchs, B. Schweighofer, H. Wegleiter. Optimization of Life Cycle Costs for Hybrid Vehicles, 11th Symposium: Hybrid and Electric Vehicles, 2014.

[8] J. Todorović, Č. Duboka, Ž. Arsenić. Operational life expectancy of rubbing elements in automotive brakes, Tribology International Vol. 28, No. 7, pp. 423-432, 1995.

5 Simulative Determination of CO_2 and Pollutant Emissions of a HEV with Multicriteria-optimised Operation Strategy in Customer Use

M. Sc. **Mark Schudeleit**, Dr.-Ing. **Marjam Eghtessad**, Prof. Dr.-Ing. **Ferit Küçükay**, Institute of Automotive Engineering (IAE), Braunschweig University of Technology, Braunschweig, Germany

Abstract

This paper describes a method to determine the real world fuel consumption as well as CO_2, NO_x and soot emissions of a HEV in simulation. To identify an optimal operation strategy (OS) for cycle and customer operation a parameter variation of the OS is conducted. The multicriteria problem of minimising various emissions is transformed into a one-criteria problem via a target function, whereby the optimum for the NEDC can be found. To simulate customer operations various vehicle measurements were carried out and analysed, which leads to a well sorted database of different driving styles, vehicle loads and driving environments. Based on this database a statistical driving environment model as well as a statistical driver model was built up in MATLAB/Simulink. This allows simulating driving operations under real customer conditions for vehicles, i.e. hybrid electric vehicles (HEV), and thus, gaining representative results for fuel consumption and emissions. It is shown that emissions in customer use differ significantly from those in legislated driving cycles.

Kurzfassung

Dieses Paper beschreibt eine Methode zur Bestimmung des realen Kraftstoffverbrauchs sowie der realen CO_2-, NO_x- und Ruß-Emissionen eines Hybridfahrzeugs mittels Simulation. Um eine optimale Betriebsstrategie für den Zyklus- sowie den Kundenbetrieb zu identifizieren, wird eine Parametervariation der Betriebsstrategieparameter durchgeführt. Das multikriterielle Optimierungsproblem, welches sich aufgrund der unterschiedlichen Emissionen ergibt, wird durch eine Zielfunktion in ein monokriterielles Problem überführt. Auf Basis der Zielfunktion lässt sich das kombinierte Optimum für den NEFZ ermitteln. Um den Kundenbetrieb in der Simulation abzubilden, wurden zahlreiche Messungen durchgeführt und analysiert. Das Ergebnis ist eine gut sortierte Datenbank für unterschiedliche Fahrertypen, Fahrumgebungen und Fahrzeugbeladungen. Mit Hilfe dieser Datenbank wurde ein statistisches Fahrer- und Fahrzeugmodell in der Simulationsumgebung MATLAB/Simulink aufgebaut. Dies erlaubt die Simulation eines Fahrzeugs, wie zum Beispiel eines Hybridfahrzeugs, unter realen Fahrbedingungen sowie die Ermittlung repräsentativer Werte für den Kraftstoffverbrauch sowie unterschiedliche Emissionen. Es wird gezeigt, dass sich die Emissionen im Kundenbetrieb deutlich von denen der gesetzlichen Zyklen unterscheiden.

1. Introduction

Increasing political pressures to reduce fuel consumption, CO_2 and pollutant emissions (i.e. soot, NO_x) moves them to the fore of further development of vehicles, such as the invention of hybrid electric vehicles (HEV). Nowadays, legislated driving cycles, like the NEDC are used to compare emission measurements. However, from a vehicle user's perspective the fuel consumption is the most relevant factor. The real world driving consumption of fuel will not be determined by legislated driving cycles, but in individual customer operations in urban, extra urban and freeway with various driving styles and vehicle loads. By now, in case of HEV no approach exists to create an operation strategy which reduces fuel consumption and various emissions in cycles and customer operation. The purpose of this study is to find a control strategy and to predict the reduction potential in combined emissions in both cycles and real driving operation.

1.1 The 3D method

The Institute of Automotive Engineering (IAE, Braunschweig, Germany) has developed the 3D method (acronym for driver, driven vehicle, driving environs) in order to take into account the real world operation of vehicles. The 3D method is a tool to determine requirements which emerge from customer operation. For instance, it can be used to forecast the representative fuel consumptions and CO_2 emissions, also considering the pollutant emissions as shown in figure 1.

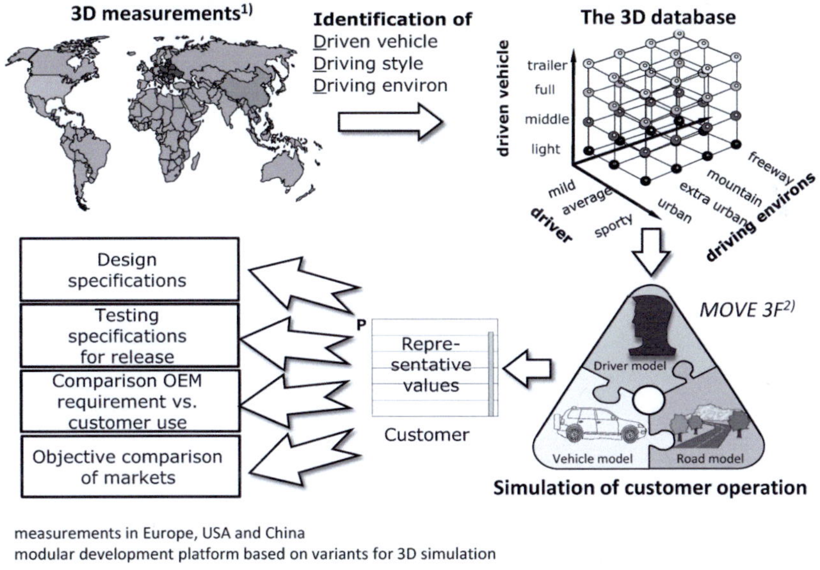

Figure 1: *Method to achieve representative results in customer operation (3D-method).*

Based on measurements in different countries collectives of probands are identified via objective criteria which leads to three distinctive groups with different driving styles. Their driving behaviour is determined by varying vehicle's load and by measurements on various representative routes. The measured data is stored in the 3D database that provides data for simulation models simulating customer operations. Depending on the specific model and the aim of the investigation simulation results can be used as design and testing specifications, benchmarks and objective comparisons of different markets.

1.2 Driving environs and driver model for customer and cycle simulation

Setting the focus on the structure of the simulation model, pictured in figure 2, the three main parts are shown which correspond to the 3D approach: the driving environs model, the driver model and the model of the driven vehicle.

The driving specifications in customer operation, such as the target speed (also known as orientation speed profile) and speed gradients are generated by a statistical orientation speed generator which uses the 3D database. The statistical driver model tries to reach this given orientation speed profile (OSP) which is a rectangle course by using the driving pedals. The driver model also implies a statistical database with vehicle, driving style and route specific actions of the driver like driving pedal gradients and its final position within a manoeuvre. [1]

In cycle simulation the driver and driving environment model are replaced by a driving guideline model which contains a driving speed controller. Its purpose is to control the accelerator and braking pedal position according to the speed difference $\Delta v = v_{desired} - v_{actual}$ in order to follow the given speed-time-course of a driving cycle.

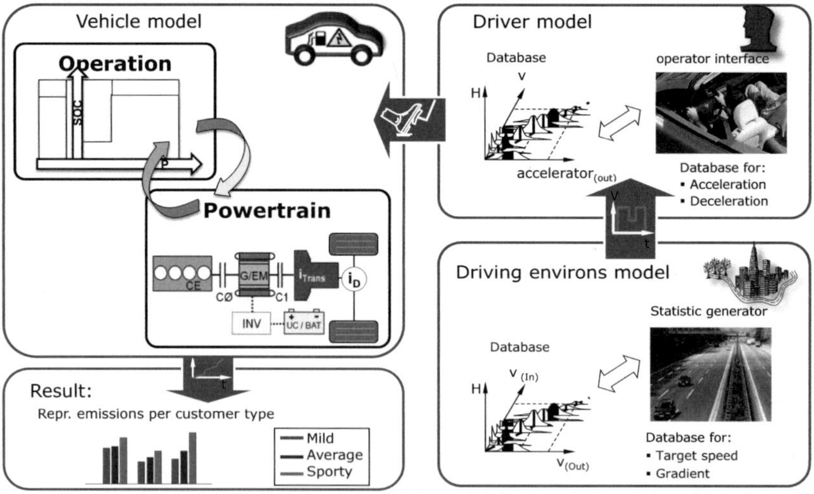

Figure 2: *Structure of MOVE 3D, a model for vehicle simulation in cycle and customer operation based on MATLAB/Simulink.*

1.3 Hybrid vehicle model

A full hybrid electric vehicle (HEV) with parallel, torque adding topology is modelled in MATLAB/Simulink. Besides an internal combustion engine (ICE) and a double clutch transmission (DCT) the drive train consists of an additional shaft between ICE and transmission which is separated by two clutches C0 and C1. The electric motor (EM) is located on this shaft.

The operation strategy of the simulated HEV chooses the actual operating mode according to the state of charge (SOC) of the battery and the driving power (P). Available hybrid operation modes are recuperation, pure electric drive, pure ICE drive and load point shift (LPS). In the latter mode an intelligent fuzzy logic controller controls through numerous rules the torque distribution between ICE and EM depending on various input parameters. Transition areas with a hysteresis are implemented between the hybrid modes to avoid alternating between them. Due to the simulated combustion engine with sufficient power the operating mode “boost” is not necessary.

2. Transient engine emission model

Focusing on emissions it is tremendously important to have sufficient emission models. Simulations demonstrates that commonly used static emission maps with the input signals of torque and speed can lead to significant differences compared to measurements. Furthermore, experience showed that in customer operation statically calculated emissions deviate even more from measurements. By adding a dynamic input signal a higher accuracy it is expected.

As a synthetic driving cycle the NEDC has less dynamic manoeuvres and furthermore, has a limited number of different driving conditions. Hence, the dynamic emission behaviour of an engine in simulation can be modelled with a rather simple empiric meta model. Neural networks as an overall model for each emission type is the simplest approach with the lowest model depth. They are fitted by measured time scribes of engine torque, engine speed and emissions in NEDC.

2.1 Empiric CO_2 emission model

Figure 3 shows simulated transient CO_2 emissions in NEDC using a neural network compared to a measurement on a test bench. From this picture follows, the simulated timely CO_2 emissions in the upper chart are approximately identically to the measurements. Furthermore, the cumulated emissions deviate just about 1 %.

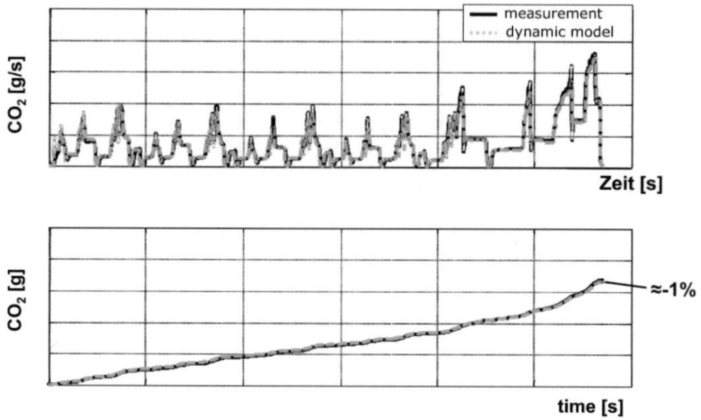

Figure 3: *Comparison of measurements and a dynamic simulation of CO_2 emissions in NEDC. For dynamic simulation neural networks with the inputs engine speed, engine torque and the torque gradient are used.*

2.2 Empiric NO_x emission model

Focusing on the NO_x emissions in figure 4, building up a meta model with neural networks leads to similar accuracy in simulation as shown before for the CO_2 emission models. The meta model simulation merely differs from the measurements in the first seconds of the NEDC when the engine is cold. Thus, with this simple approach it is not possible to achieve a high accuracy during cold-start phase of an engine. However, the behaviour after warm-up and the similar cumulated emissions imply the appropriateness of the simulation model.

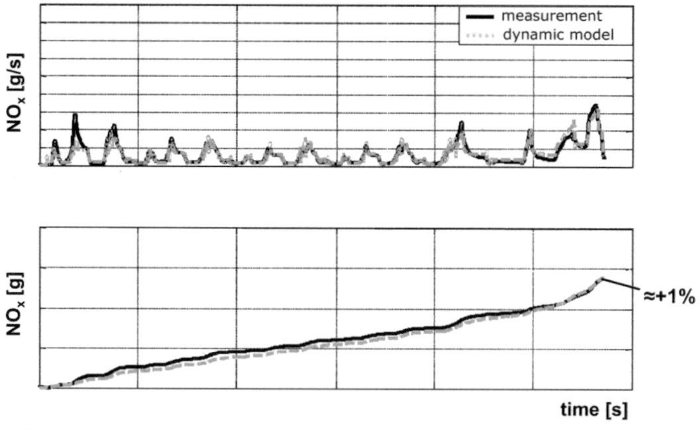

Figure 4: *Comparison of measurements and a dynamic simulation of NO_x emissions in NEDC. For dynamic simulation neural networks with the inputs engine speed, engine torque and the torque gradient are used.*

2.3 Empiric soot emission model

From figure 5, illustrating the soot emissions of simulation and measurement in NEDC, it can be recognized that the simulation differs just in the peaks of soot emissions which are not reproducible as they appear differently for same NEDC manoeuvres. However, the cumulated emissions of the simulation deviate from the measurements less than 1%. Therefore, it may be concluded that the actual meta model is appropriate for dynamic NEDC simulation.

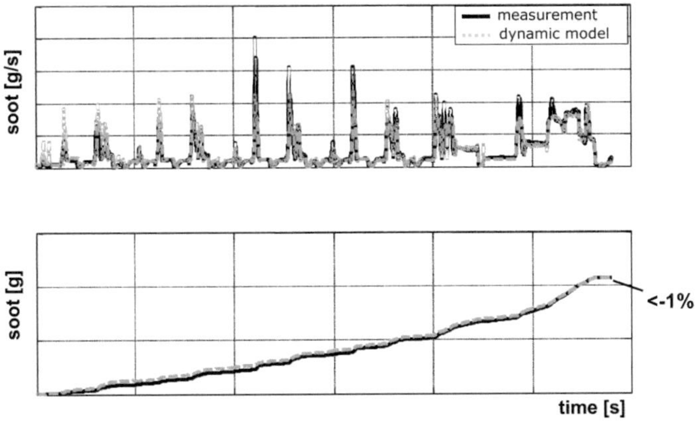

Figure 5: *Comparison of measurements and a dynamic simulation of soot emissions in NEDC. For dynamic simulation neural networks with the inputs engine speed, engine torque and the torque gradient are used.*

2.4 Semi-empiric dynamic emission model for customer operation

Modelling dynamic emissions of customer operation needs a more complex approach as the connections between manoeuvres and emissions are more complex. For this reason the transient behaviour is modelled by semi-empiric meta models which were created after analysing representative customer driving manoeuvres using the 3D method. It results that the emission meta model for customer simulation are based on a time-variant transfer function (first and second order). The inputs are a stationary amount from a quasi-static engine map and a transient time constant which is modelled with a neural network. This enables more exact calculations of real emissions of engine in customer operation.

3. Multicriteria Optimisation of operation strategy

It is known that the operation points of the ICE differ significantly in customer operation depending on the driving style and the driving environment. This also has effects on emissions and fuel consumption. In order to minimise both in customer operation it is insufficient just taking into account operations in driving cycles. Thus, it is obligatory that the HEV achieves low emissions and fuel consumption in both cycle and customer operation by optimising the operation strategy (OS). The OS must lead to an optimal compromise between various emissions which partly have a contrary behaviour. Thereby, it is a multicriteria optimisation problem. Furthermore, in cycle and customer operation the optimized OS should show similar results regarding average emissions. Although, the distribution of emissions in different driving styles in customer use has to be reduced. In general, changes of influences, also known as *basic parameters*, must have a low sensitivity concerning increases in consumption and emissions.

In order to fulfil the criteria, minimising emissions and consumption as well as robustness, investigations are conducted by varying parameters of the operation strategy and identifying the effects on emissions. Based on the results the optimised robust operation strategy can be determined by analysing a target function.

The Target function summarises the target parameters CO_2, NO_x and soot for each parameter set. In the first step, the target parameter are normalised to an interval [0:1] by dividing them through their maximum to get dimensionless variables with the same relative magnitude. Finally the target function Z arises by the weighted addition of the previously normalised criteria using weighting factors for each target parameter as shown in formula (1) beneath.

$$Z=g_1 \cdot CO_2+g_2 \cdot NO_x+g_3 \cdot soot+\ldots \qquad (1)$$

weighting factors: $g_1, \ldots, g_n$, n=number of criteria

The simulation results of all parameter sets lead to a one multi-dimensional matrix spread out by the variation parameters and the value of the target function Z. On the basis of this a robust operation strategy is identified.

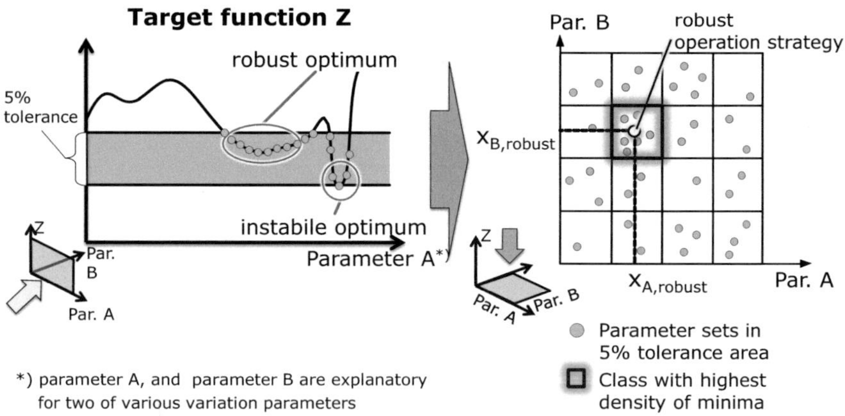

Figure 6: *Left: Identification of parameter set with optimum close target function (5% tolerance).*
Right: Classification of 5%-optimum parameter sets to identify a robust operation strategy (OS) being in the class with the highest density (explanatory example).

To find a robust optimum not only the minimum but also the parameter sets which with similar results of the target function (up to 5% tolerance) are classified like in figure 6 (right). A robust optimum which is defined as minimum surrounded by a flat area has the highest density of minimal and nearly minimal points. Thus, in figure 6 (right) the grid with the greatest amount of points is chosen. The size of the grids is estimated by the number of identified parameter sets in the 5% tolerance using the Sturges rule [2].

4. Results in cycle and customer operation

Figure 7 shows the result regarding fuel consumption (equivalent to CO_2 emissions), NO_x and soot emissions of a HEV with robust operation strategy in customer and cycle operation as well as the emissions of a conventional vehicle in NEDC. The hybrid drive causes clearly less overall emissions as the conventional vehicle. For HEV simulation the results in NEDC also differ significantly from the ones in customer operation. In this driving cycle the HEV leads to much lower fuel consumption than expected in real use. The NO_x emissions of a HEV in NEDC are similar to one mild and several average drivers, whilst the sporty driver produces considerably higher NO_x emissions. In contrast to customer operation the soot emissions are comparably high in cycle use. as emissions in real operation differ drastically even in cases of a HEV with a robust operation strategy all these results imply that it is insufficient solely look on cycle emissions.

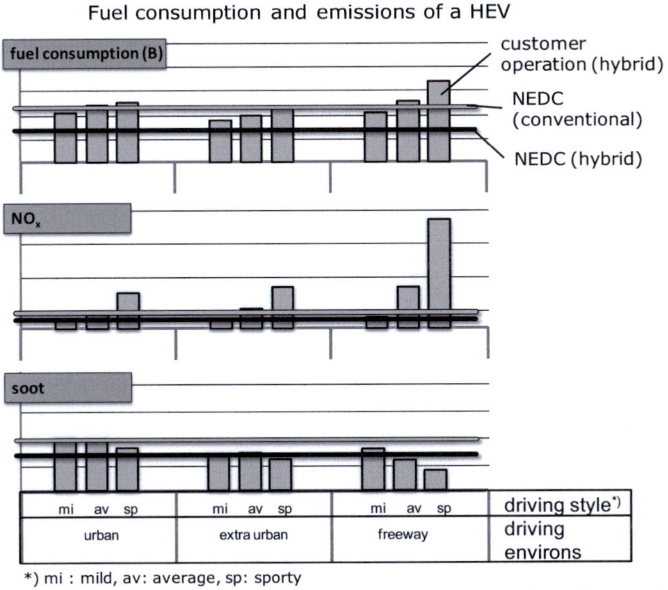

Figure 7: *Fuel consumption, NO_x and soot emissions of a HEV with robust operation strategy in customer operation compared to NEDC and a conventional vehicle in NEDC.*

5. Conclusion

In this paper an approach was proposed to build up empiric and semi-empiric meta model for dynamic engine emissions (CO_2, NO_x, soot) in order to gain more accurate simulation results. Furthermore, it was shown the process to optimise the HEV operation strategy not only for cycle but also for customer use cases regarding multiple criteria like various emissions also considering the robustness which is defined as a stable behaviour due to parameter changes. The identified operation strategy should be especially advantageous in customer operation as it leads to similar emissions. IT was identified by building a target function Z and taking the best results. The stable area was located where the most of these results were located. Further investigations showed that the results in fuel consumption (CO_2), NO_x and soot in NEDC differ significantly from those in customer operation. This fact shows that it is insufficient solely to focus on driving cycles as testing procedure.

References

[1] Müller-Kose, J. P.: Repräsentative Lastkollektive für Fahrzeuggetriebe (Representative Load Spectra for Vehicle Gearboxes), PhD thesis, Institute of Automotive Engineering, Braunschweig University of Technology, Shaker, Aachen, 2002.

[2] Sturges, H. A.: The choice of a class interval. In: Journal of the American Statistical Association. 21 (1926), Nr. 153, S. 65-66.

Abbreviations

Symbol	Meaning
EM	Electric motor
HEV	Hybrid electric vehicle
ICE	Internal combustion engine
LPS	Load point shift
mot	motoric
NEDC	New European Driving Cycle
OPS	Operation point share
opt	optimal
OS	Operation strategy
recup	recuperation

6 Estimation of the Optimal Length of Run of Electromobile Taking Into Account its Main Parameters

Prof. **Olexandr Kuzemin,** Prof. **Mykola Slipchenko,** Prof. **Viktor Pismenetskyi,** Dipl.-Ing. **Mykhailo Gurtovyi,** Kharkiv National University of Radio Electronics, Kharkiv, Ukraine

Abstract

Simulation of electromobile (EM) regimes functioning makes it possible to estimate such important parameters as length of run, acceleration, accumulator battery (AB) capacity and type. At present the analytical dependences of the EM length of run on the AB discharge depth are practically absent. That is why the problems of analysis and modeling of EM parameters and characteristics are of significance and practical interest for designers and producers of automobiles.

The computational experiment with further analysis of the experimental data has been carried out on basis of the length of run algorithm simulation. Analysis and modeling of the main EM parameters has made it possible to receive concrete results of the depth of discharge dependences on length of run for different weight of effective load.

Research of length of run dependence on the AB capacity and speed of the EM motion has been carried out with the regression method. The search for the optimal values of the EM motion speed and the AB capacity by criterion of the length of run maximum according to the gradient descent is presented in this paper. The optimal values of the motion speed for the EM with the mass equal to 1,4 ton have been received (30 km/h and 85 Ah) as a result of two-parametric optimization.

1. Introduction

Now a market of electromobiles (EM), infrastructures of their sale and service are forming. Due to fact that reduction of emissions of carbonic gas is one of the main advantages of EM in comparison with vehicles with an internal combustion engine. Besides, intracity use of modern electromobiles with limited maximum speed and at rather high quality of roads allows to lower requirements on a number of constructive characteristics of assemblies and units that creates preconditions for their certain simplification in comparison with the usual automobile. Low cruising range and high cost of EM are the main lacks limiting their applicability. However, since 2014 in the USA it is possible to produce either hybrid cars or electrocars. So, the EM of innovative concern Teslamotors passes 335 - 425 km on one refuelling, and the maximum speed is 200 km/h [1]. Due to increasing prevalence of the electrotransport, simulation of the EM regimes functioning is an important task of automotive industry, because it makes possible to estimate such important parameters as length of run, acceleration, AB capacity and type.

2. The analysis of literary data and problem statement

The estimation of energy consumption of an EM depends on numerous parametres, such as torque of the motor, speed of movement, weight, current and temperature of accumullator battery. Therefore a mathematical model of an EM is necessary for definition of potential possibilities of energy power unit concerning interaction of its other components. It is known, that EM's length of run depends on it energy consumption and capacity of an AB [2]. Now analytical dependences of length of run of an EM on depth of discharge of an AB and speed of movement practically are absent. Therefore the problem of the analysis and modelling of parametres and EM's characteristics is important and represents practical interest.

3. The aim and research problems

The aim of this work is definition of dependence of length of run of an EM on depth of the discharge of the AB of lead-acid type in a standard driving cycle on a city at a variation of AB's capacity. The main research problems are:

- definition of dependence of length of run of an EM on depth of discharge of the AB for different payload weight;
- research of an effect of AB's capacity (in A·h) on the length of run at movement with constant speed;
- definition of optimum values of parametres (AB's capacity and speed of movement).

4. Main results

The process of the electric motorcar acceleration up to the normal speed depends on the following main characteristics: the EM's weight, the acceleration time, the frame aerodynamic characteristics, the electric motor power, the AB peak current, the capacitor starting capacitance [3].
In order to start moving of an EM the tractive effort force of drive wheels F_{te} must exceed the sum of other forces (fig.1):

$$F_{te} = F_{rr} + F_{ad} + F_{hc} + F_{la} + F_{\omega a}, \qquad (1)$$

where F_{rr}- rolling resistance force; F_{ad} - aerodynamic drag; F_{hc} - hill climbing force; F_{la}- linear acceleration force; $F_{\omega a}$ - angular acceleration force [3 - 6].
The energy demanded for movement every second, is expressed by power P_{te}:

$$P_{te} = F_{te} \times v, \qquad (2)$$

where P_{te} - power, demanded for movement every second, W·s; v – speed, m/s.

The inefficiencies of a motor, a controller and a gear system mean that the motor's power is not the same as the traction power, and electrical power required by the motor is greater than mechanical output power, according to the equations: (3):

$$P_{mot\_in} = \frac{P_{mot\_out}}{\eta_m}, \; P_{mot\_out} = \frac{P_{te}}{\eta_g}, \tag{3}$$

where P_{mot_in} –electric input power, W; P_{mot_out} – mechanical output power, W; η_m - efficiency of the motor and its controller; η_g - efficiency of the gear system.

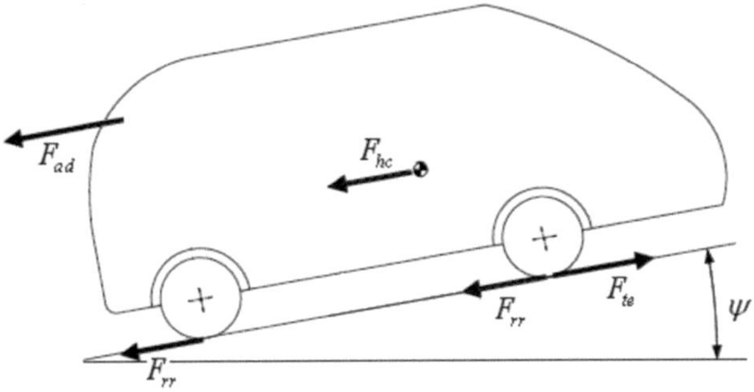

Figure 1: *The forces acting on a vehicle*

Equations (3) are correct in the case where the EM is being driven. However, it is necessary to apply following equations (4) when EM brakes:

$$P_{mot\_in} = P_{mot\_out} \times \eta_m, \; P_{mot\_out} = P_{te} \times \eta_g, \tag{4}$$

Equations for definition of AB's charge removed and charge supplied look like:

$$CR_{n+1} = CR_n - \frac{\delta t \times I^k}{3600}, CS_{n+1} = CS_n + \frac{\delta t \times I}{3600}. \tag{5}$$

where CR_n – charge removed from the battery by the nth step of the simulation, A·h; CS_n - charge supplied from the battery by the nth step of the simulation, A·h; k – Peukert coefficient; δt - time, s; I - discharge current, A [7].
The depth of discharge (DOD_n) of AB is the ratio of the charge removed to the original capacity C_P, as it is specified in [4, 7]:

$$DOD_n = \frac{CR_n}{C_P}, \tag{6}$$

where C_P - Peukert Capacity, A·h.

The electric motor, energy storage device (AB or hybrid system with supercapacitors, fuel elements, or flywheels) and mechanical drive are among main components of simulation [8]. The cycle SFUDS has been chosen as a standard driving cycle within a city and it is shown on fig. 2 a. The schedule of power consumption of given cycle is shown in fig.2 b according to work [6].

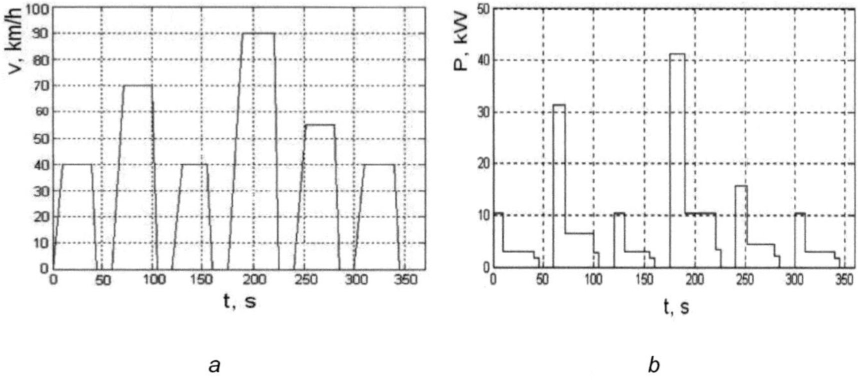

Figure 2: *Standard driving cycle SFUDS: a – graph of speed against time; b - graph of power against time*

Further, applying data from table 1, the computing experiment, which basis is the algorithm of modelling of length of run d, has been made (fig. 3).

Table 1: *Technical data of the intracity passenger EM*

Weight without payload, t	Front area, m^2	Power of the additional electric power consumers, W	Friction factor/ Aerodynamic drag coefficient	Efficiency of the gear system η_g
1,4	1,8	250	0,19 / 0,31	0,95

Generally in this algorithm AB's depth of discharge proceeding from power given to load and recuperation at braking is calculated. By means of algorithm, which is presented at fig.3, length of run d has been defined at movement in a city cycle for various payloads (fig.4 a). Dependence of AB's depth of discharge has linear character and length of run decreases with growth of weight of a payload.

Dependence of lenght of run d from speed v with variation of the lead-acid type AB's capacity C, constructed on the basis of array of source data (7) is presented at the fig.4 b. In this case EM's weight in the equipped condition changed slightly. As follows from fig. 4 b, optimum speed of movement is equal 30 km/h. Position of maximum of length of run is connected with presence of a certain operational torque of the electric motor.

Dependences d=f (v) and d=f (C) for the array of source data are resulted on fig. 5.

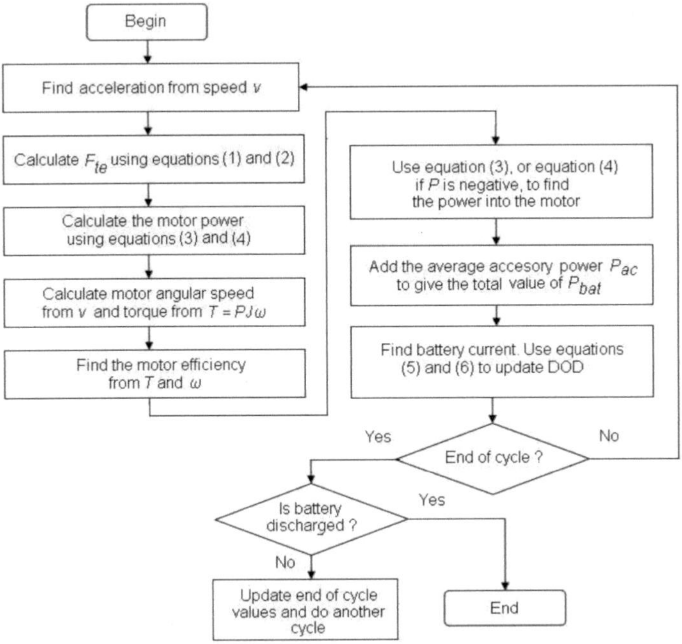

Figure 3: *Length of run simulation algorithm*

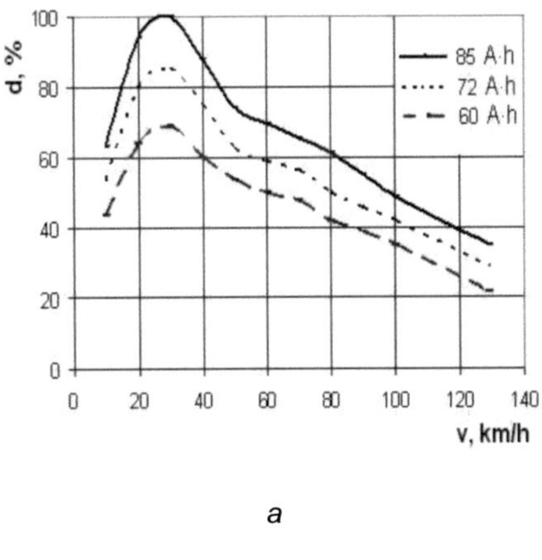

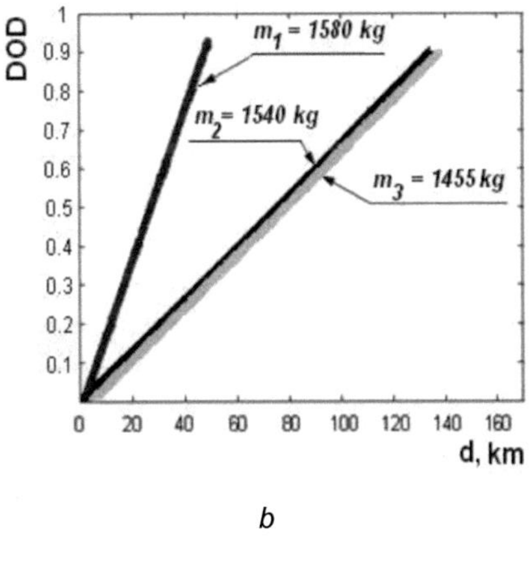

Figure 4: *Length of run: a – dependance of length of run d on speed v; b - dependance of battery DOD on length of run d for different payload weight in standart driving cycle*

$$ASD=\begin{cases}
0 & 43 & 45 & 50 & 55 & 60 & 65 & 72 & 75 & 80 & 85 \\
10 & 30 & 31.5 & 37 & 40 & 43.75 & 48.5 & 54 & 56.25 & 61.25 & 65.25 \\
20 & 45 & 47 & 53 & 60.5 & 63.25 & 70.5 & 79.75 & 80 & 88.75 & 96.25 \\
30 & 47.5 & 50 & 56.25 & 62.5 & 69.5 & 75 & 84.5 & 88 & 95 & 100 \\
40 & 39.5 & 41.5 & 49.25 & 55 & 62.25 & 68.75 & 75.75 & 80 & 87.5 & 93.5 \\
50 & 39.25 & 40.5 & 45 & 49.25 & 53.25 & 57.5 & 63 & 66.25 & 70 & 75 \\
60 & 35.5 & 38.75 & 43.25 & 46 & 51.75 & 54.25 & 59.25 & 62.5 & 66.75 & 71.25 \\
70 & 34.25 & 36.75 & 39.25 & 42.5 & 48.25 & 51.75 & 56.5 & 61.25 & 63.75 & 67.25 \\
80 & 30 & 31.75 & 34.5 & 39 & 43 & 45.75 & 50 & 54.5 & 58.75 & 63.75 \\
90 & 26.75 & 29.25 & 32 & 36.5 & 37.5 & 42.5 & 46 & 48.75 & 53.25 & 57.75 \\
100 & 24.25 & 24.5 & 29.25 & 32.5 & 34.5 & 37 & 41.25 & 44.5 & 46 & 49.25 \\
129 & 16 & 18.75 & 19.25 & 20 & 22 & 25 & 28.25 & 31.25 & 32.5 & 34.5
\end{cases} \quad (7)$$

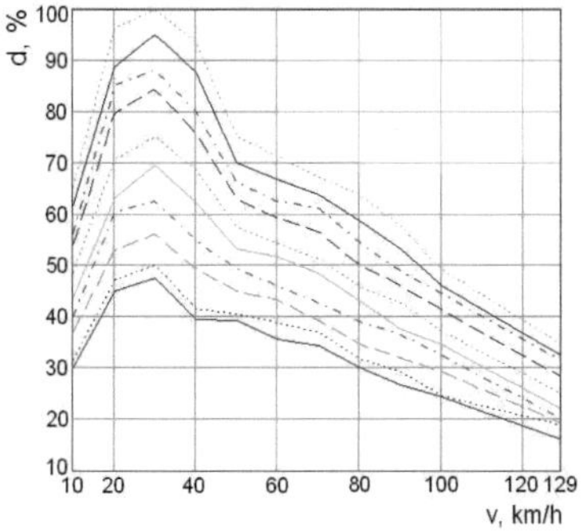

a

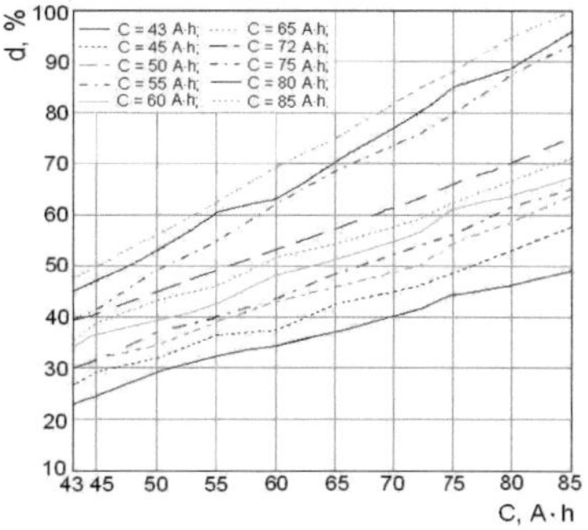

b

Figure 5: *Schedules of dependences for initial data: a – dependance of d on v; b - dependance of d on C*

The research of dependence d = f(C,v) was carried out by equation of a regression of the fifth order:

$$\bar{f}_x = 4.626 - 1{,}41x + (4.54\cdot10^{-2})x^2 + (-4.98\cdot10^{-4})x^3 + (1.76\cdot10^{-6})x^4 + \\ + (1.42\cdot10^{-13})x^5 + (1.56\cdot10^{-1})y + (9.44\cdot10^{-2})xy + (-2.63\cdot10^{-3})x^2y + \\ + (2.3\cdot10^{-5})x^3y + (-8.84\cdot10^{-8})x^4y + (-9.87\cdot10^{-15})x^5y, \quad (8)$$

where *x* – speed v, km/h; *y* – AB's capacity C, A·h. As can be seen from the table 2, the probability that the second degree equation erroneously equals to 4,58%, and for fifth degree equation it is equal to 1,23%.

Table 2: *Errors of mathematical model*

Degree of equation	1	2	3	4	5
Probability of an error, %	8,58	4,58	1,39	0,41	1,23

Further by means of generalised techniques of two-parametric (parametres x and y) optimisation by a method of gradient descent are defined optimum values of capacity C and speeds v. The algorithm of the given method consists of following stages [9]:
1. Input of initial values of parametres (x_0, y_0), their range of definition (x_{min}, y_{min}, x_{max}, y_{max}), discretenesses of parametres (h_x, h_y), precisions of calculation ε.
2. Definition of derivatives: $f'_{x_i}(x_i, y_i), f'_{y_i}(x_i, y_i)$.
3. Calculation of the modulus of the gradient:

$$r = \sqrt{\left[f'_{x_i}(x_i, y_i)\right]^2 + \left[f'_{y_i}(x_i, y_i)\right]^2}\ . \quad (9)$$

4. Calculation of a directing sine and cosine of gradient:

$$\cos\alpha = \frac{f'_{x_i}(x_i, y_i)}{r},\ \sin\alpha = \frac{f'_{y_i}(x_i, y_i)}{r}. \quad (10)$$

5. Calculation of coordinates of the following point of search:

$$x_{i+1} = x_i + h_x \cos\alpha,\ y_{i+1} = y_i + h_y \sin\alpha. \quad (11)$$

6. Condition check of achievement of an optimum point on parametre ε.
7. Verify parametres-coordinates in the field of admissible values.
8. Output of an optimum parametres-coordinates.

The dynamics of an automatic scan of the optimum parametres of AB's capacity C and speed v by means of gradient descent by regression equation (8) is represented at fig.6.

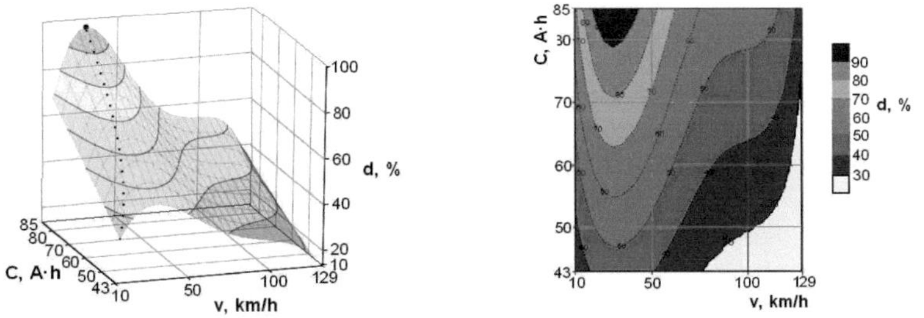

a *b*

Figure 6: *Dependence d=f(C, v): a – search of the optimum parameters C and v; b - dependance of C on v*

As can be seen from fig.6, a, the line of search of an optimum point (black point) is built on a surface d = f(C,v) by two-parametric optimisation. The projection of a trajectory of search of an optimum point to plane C-v is shown in fig.6, b.
Then trajectories of search of a point with optimum coordinates, which are shown in fig.7, have been drawn.

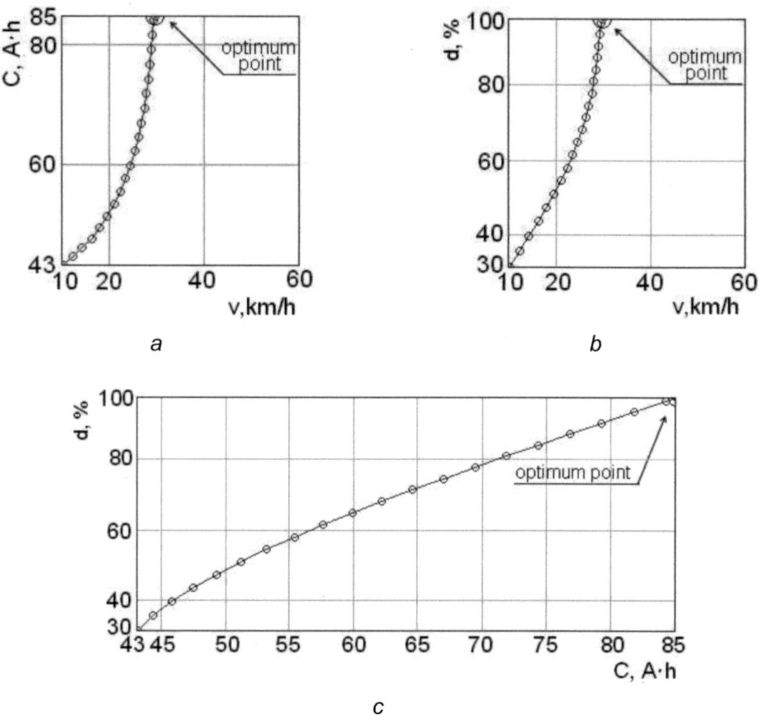

Figure 7: *Trajectories of search of an optimum point: a - dependance of C on v; b - dependance of d on v; c - dependance of d on C*

Projections of a trajectory of search of an optimum point to planes C-v, d-v and d-C allow to define a necessary amount of iterations for its determination. From the resulted graphs follows, that an optimum value of parametres C and v by criterion of maximum of parametre d are 85 A·h and 30 km/h accordingly.

5. Conclusions

1. The analysis and modelling of key parametres of EM have allowed to receive concrete outcomes of dependances of depth of discharge from length of run for

various weight of a payload. It is shown, that length of run decreases with magnification of a weight of useful payload in a standard driving cycle on a city.
2. Regression model, which connects AB's capacity and EM's speed has been determined. Search of optimum values of a EV's speed of movement and AB's capacities by criterion of a maximum of length of run according to gradient descent has given an optimum value of speed of movement, equal 30 km/h that agree with outcomes presented in works [2, 10].

References

[1] America transfers to electric vehicles. Economics and Business. Available at : http:// www.newsland.com/news/detail/id/1229354/ (accessed 25 February 2014)

[2] Giessler, M., Fritz, A., Paul, J., Sander, O., Gauterin, F., Műller-Glaser Klaus-D. (2012). Converted vehicle for battery electric drive: Aspects on the design of the software-driven vehicle control unit. B. Baker, L. Morawietz (Eds.), proceedings of the 2nd EEVC (pp. 85-97). Dresden: Expert verlag.

[3] Efremov, I. S., Prolyguin, A. P., Andreev, Yu.M., Mindlin, A.B. (1984). Theory an calculation of the electric motorcar draught drive. *Theory and calculation of the electric motorcar draught drive* (pp. 68 - 83). Moscow, CT: Vysshaya shkola.

[4] Larminie, J., Lowry, J. (2003). Electric Vehicle Techology Explained. *Electric Vehicle Modelling* (pp. 183 – 196). Chichester, CT: John Wiley & Sons Ltd.

[5] Guzzella, L., Sciarretta, A. (2005). Vehicle propulsion systems. Introduction to modelling and optimization. *Models of Electric and Hybrid-Electric Propulsion Vehicle Systems* (pp. 68 – 101). Berlin, CT: Springer-Verlag.

[6] Dhameja, S. (2002). Electric Vehicles Battery Systems. *Electric Vehicle Battery Capacity* (pp. 43 – 53). Boston, CT: Newnes.

[7] www.elkraft.ntnu.no/eno/Papers2009/EVS24_final_paper-giuseppe.pdf

[8] Anosov, V. N., Kaveshnikov, V. M., Orel, A. V. (2008). Analysis of changes in the discharge capacity of the accumulator battery. *Transport: science, technique, control, 6,* 33 – 36.

[9] Lvovich, Ya. E., Frolov, V.N. (1986). Theoretical bases of designing, technology and reliability of radio-electronic equipment. *Processing of the experimental data by regression analysis* (pp. 79-84). Moscow, CT: Radio and svyaz.

[10] Busyguin, B. P. (1979). Electric motorcars (Calculation methods). *Evaluation of electric vehicles and the analysis of their profiles* (pp.52-67). Moscow, CT : MADI.

7 Driver Individual Energy Consumption Forecast

Dipl.-Inform. **Tobias Bär,** Dr.-Ing. **Dennis Nienhüser,** Prof. Dr.-Ing. **J. Marius Zöllner**, FZI Research Center for Information Technology at the Karlsruhe Institute of Technology, Karlsruhe, Germany

Abstract

An accurate energy consumption forecast is crucial to accomplish the turn to electric vehicles. Especially the fear of breaking down due to running out of power discourages people purchasing electric cars. An accurate driving range prediction increases the usable driving range since less energy reserves are needed. In most cases, the energy consumption is computed by means of map and vehicle information only. The individual driving behavior of the motorist is barley considered. However, the habitual speed the driver chooses, or his braking and acceleration habits have a major impact on the energy usage.

In this work, we create a driver profile of the individual driver whilst driving and improve the energy estimation forecast thereby. The driver profile consists of the driving-style and habitual parameters, like the habitual highway speed or the commonly used acceleration. With regards to the parameters gathered in this profile, a driver-specific speed profile for the planned track or the track ahead respectively, is predicted.

The predicted speed profile, together with the height profile of the track, on-line traffic information, and a vehicle model leads to an improved, driver-specific energy forecast.

1. Introduction

The results of a survey carried out in Germany show that every seventh car owner is willing to buy an electric vehicle before 2020 [4, 5]. Today, two major impediments inhibit the demand. These are the high acquisition costs and the lack of extensively available infra structure. The latter is further worsened by the limited cruising range of today's electronic cars.

One opportunity to get past the fear of breaking down because of the electric car running out of power is providing an accurate and trustworthy prediction of the cruising range which is more or less equivalent to an accurate energy consumption estimation. However, an accurate prediction of the energy consumption needed for a particular trip has to be well adjusted to the particular driver, since the driving behavior is an important impact factor on the energy spent.

In experiments, made in our driving simulator, we observed variances in energy consumption under exactly the same circumstances up to 25 percent by letting participants drive a 9.7 kilometer long test track, modeled like a German country road [1]. Other research groups, mainly researching on energy saving applications, gathered comparable results [3, 7, 8].

However, the driver's impact is not easy to consider within the energy consumption forecasting process. The same driver may show different behavior depending on his

mood, his mental stress level, his familiarity with the given traffic situations, etc., with all these parameters being mainly immeasurable.
The focus of this work is the determination of a driver individual and driver-characterizing driver profile, holding habitual properties describing the driver in interaction with his vehicle and his environment. Subsequently, this driver profile is used to driver-individually forecast the energy consumption of the track ahead.

2. Implementation

In this work, the process of forecasting the energy consumption on a given track is accomplished by forecasting a speed-profile on a given track. The speed-profile is deduced by successively arranging predicted driver's actions in a given traffic situation. Within the traffic situation approaching a stop sign the predicted actions would be {maintain speed} followed by {coast} followed by {brake}. However, the duration and the parameterization of the actions are depending on the driver's driving profile. Economic drivers, for instance, will coast as long as possible, whereas racy drivers would rather brake and maintain their speed for a longer time.

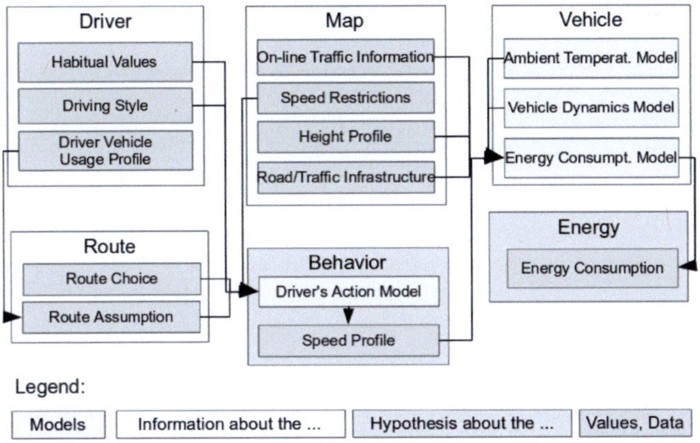

Figure 1: *Our implementation estimates the energy used for a given track by means of forecasting a speed profile. The speed profile depends on various information sources, e.g. the driving-style and the habitual parameters of the driver.*

Figure 1 shows the components influencing the energy consumption forecast. This work focuses on the driver's impact, expressed by his driving style and the estimated habitual parameters.

2.1 Driver Profile

The driver profile consists, among other values, of the driver's habitual parameters and the driving style of the driver. It is estimated in a probabilistic manner evaluating a particular traffic scene by means of fuzzy logic [2].
The habitual values are calculated in a context-sensitive manner, too.

For example the habitual velocity $v_h(R)$ is calculated for every speed restriction R, provided by OpenStreetMap data. The habitual velocity $v_h(R)$ is calculated by means of averaging the last N = 1500 speed measurements when the driver was driving unaffected by other road users or the road curvature.
More precisely, the driver habitual value for the velocity is expressed as the average velocity for a given velocity restriction R and the standard deviation to it.
To determine if the driver was impaired by other road users the laser-scanner data is analyzed. Other habitual values stored in the driver profile are the habitual accelerations, decelerations, or the safety distance in particular traffic situations, like stop-and-go or approaching stop signs or traffic lights.

2.2 Driver's Actions and Speed-Profile Calculation

To calculate the speed-profile for a given track a micro-simulation of the vehicle state is carried out which virtually performs the driver's actions [6, 9]. The actions of the driver are listed in table 2.1. All actions are implicitly influenced in their duration by the target speed, which is set to the driver's habitual velocity within the target speed restriction.

Action	Parametrized by
Brake	Habitual Brake Deceleration Habitual Velocity
Coast	Habitual Velocity
Maintain Speed	Habitual Velocity
Accelerate	Habitual Acceleration Habitual Velocity

Table 1: *Drivers Actions and their parameterization by the Driver Habitual Parameters*

These driver-individually parameterized actions are virtually performed one after another on the given track and lead to a speed profile forecast of the given track ahead. The speed forecast is further influenced by the on-line traffic information and the curve radii of the track.

2.3 Energy Forecasting Using the Speed Profile

The energy consumption is calculated by means of an energy consumption model processing the speed profile, the tracks height profile, and the vehicle's energy-consumption characteristic as inputs. The energy consumption model can emulate an electric vehicle characteristics and a gasoline engine modeling an Audi Q5.
The energy forecast is implicitly influenced by the habitual velocity and the habitual acceleration. Consider a simple scenario with two different drivers A and B: We estimate the consumption of a acceleration from standstill to a given velocity restriction R. Since the forecasted speed profile would be depended on the individual acceleration $a_{hA,B}$ and the individual velocity $v_{hA,B}(R)$, the forecasted speed profile would be different and thus, the estimated energy consumption would be different, too.

3. Experiments and Test Setup

Our approach is validated in two different test scenarios. In section 3.2 the energy consumption of a 25 km highway drive is forecasted. The experiments show, that the forecast is getting more and more accurate whilst driving.
In section 3.3, the energy forecasts of two drivers riding the our 'Turmberg' test track are compared. Prior to that, in section 3.1, our test vehicle is introduced briefly.

3.1 Test Setup and Test Vehicle

Our test vehicle CoCar (Cognitive Car) is a modified Audi Q5 containing measurement equipment to record CAN-bus, front-camera, gaze tracking, IMU, and laser-scanner data synchronously. The energy consumption of the car is provided on the CAN-bus. The Robot Operating System (ROS) [10] is used to record, replay and process the sensor data. On-line traffic information, provided by Google Maps, can be stored to reconstruct the traffic conditions for the off-line evaluation. Figure 2 shows our test car and the test track with its height profile of the Turmberg Drive experiments (Section 3.3).

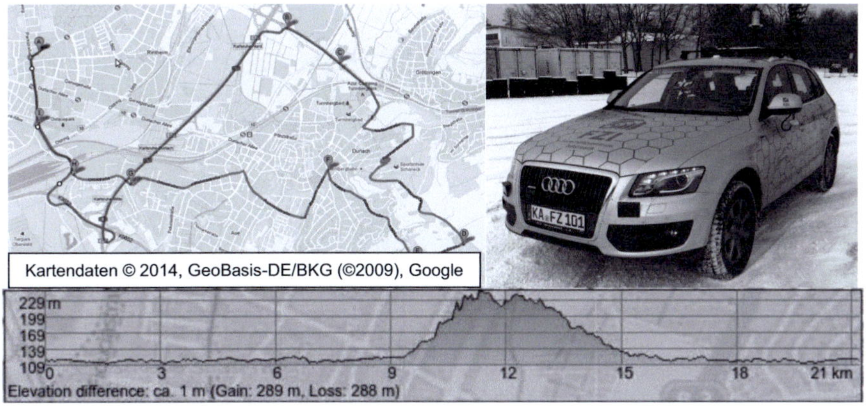

Figure 2: *The test track of the Turmberg Drive (section 3.3) and the test vehicle CoCar used for the evaluation and the determination of the habitual driver parameters.*

3.2 Highway Drive

The accuracy of the energy consumption forecast by means of individual velocity and acceleration was tested on an 18 km highway drive from Karlsruhe to Rastatt and back. The test drive includes the drive on the feeder road (velocity restriction 60 km/h and 80 km/h), and the highway itself with an velocity restriction of 120 km/h and unlimited in the end. Figure 3 shows the energy consumption estimation at the start (left) and the end of the trip (right). At the start of the trip, the habitual parameters are uninitialized. Hence, the individual energy consumption forecast (magenta) is similar to the default forecast (blue). At the end of the trip, the habitual values of the driver are estimated, improving the accuracy of the energy forecast. The measured energy consumption (ground truth) is plotted in red.

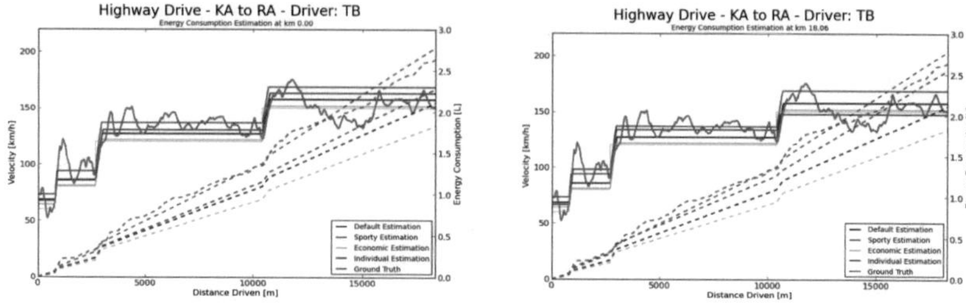

Figure 3: *Real and forecasted energy consumption on a highway drive from Karlsruhe to Rastatt. The dashed line visualizes the energy consumption; the solid lines visualize the velocity. The red lines (ground truth) are measured by reading the CAN-bus, all others lines are prognoses about the driver behavior and the energy consumption respectively.*

Figure 4 (left) shows the evolution of the energy forecast accuracy whilst driving. It can be seen, that the energy consumption forecast gets the more and more precise, the more and more the driver travels. Figure 4 (middle) shows the habitual velocity parameters at the end of the highway drive.

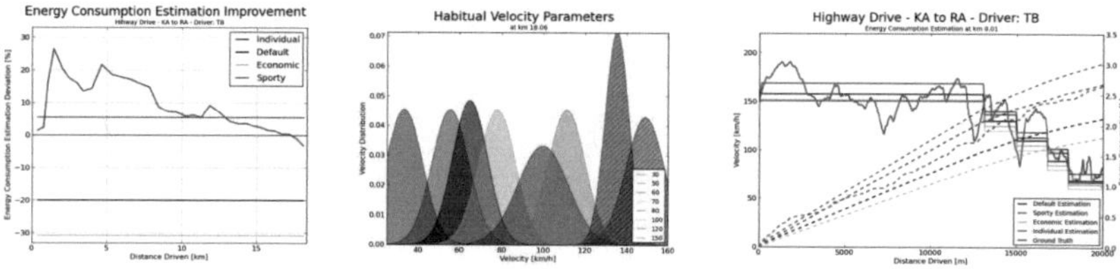

Figure 4: *Improvement of the energy consumption forecast (left). The estimated habitual velocity parameters influencing the forecast (middle).*

After travelling from Karlsruhe to Rastatt, an energy consumption forecast is made for the return trip (left plot of figure 4). Although, there are different velocity restrictions and traffic conditions, the individual error consumption of the return trip can be estimated more accurate using the driver habitual parameters (error of 3.8 %, whereas the sporty estimation forecasts with an error of 15.3% compared to ground truth).

3.3 Turmberg Drive (~25 km)

In our second experiment, we compare two different test drivers on a ~25 km long trip over the Turmberg (Karlsruhe, Durlach). The test track is chosen because of its variety. It consists of a highway drive, a drive through rural areas, a drive up and down the hill, and crowded city traffic. The height profile and the test track can be seen in figure 2.

Table 2 compares the two drives. The energy consumption estimation is made with the driver's own habitual parameters and the habitual parameters of the opponent driver.

	Drive A (TB)	Drive B (DN)
Habitual Values of Driver TB	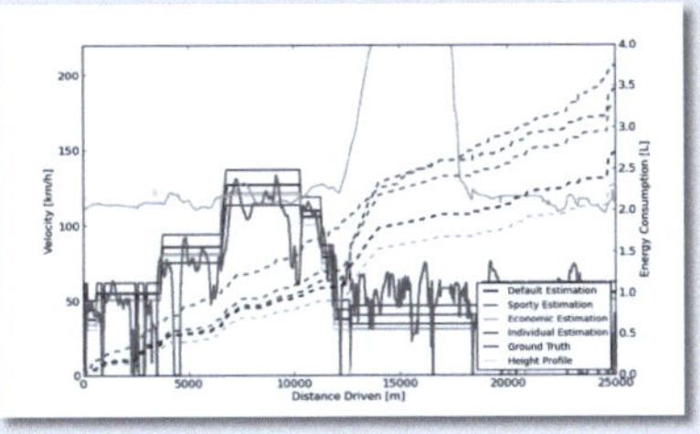	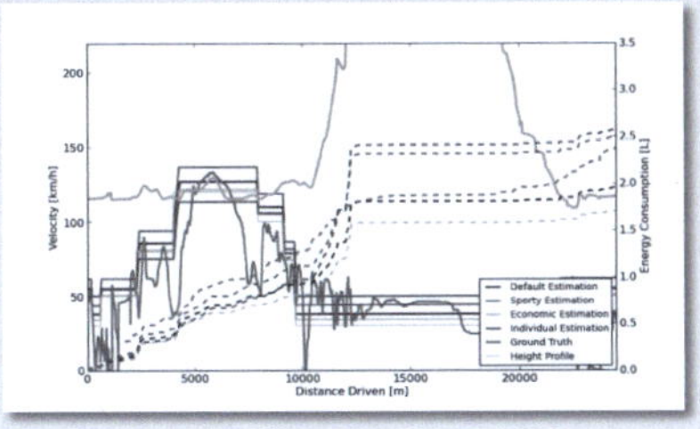
Habitual Values of Driver DN	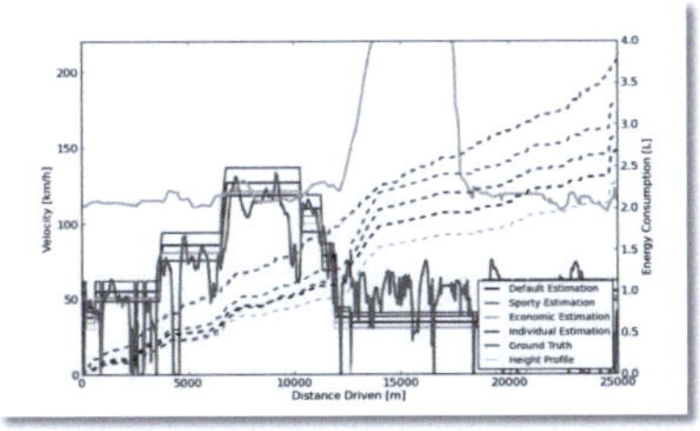	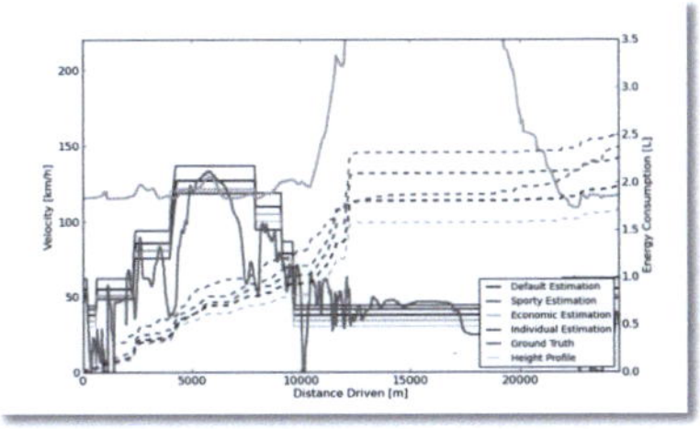
Comparison Driver A and B	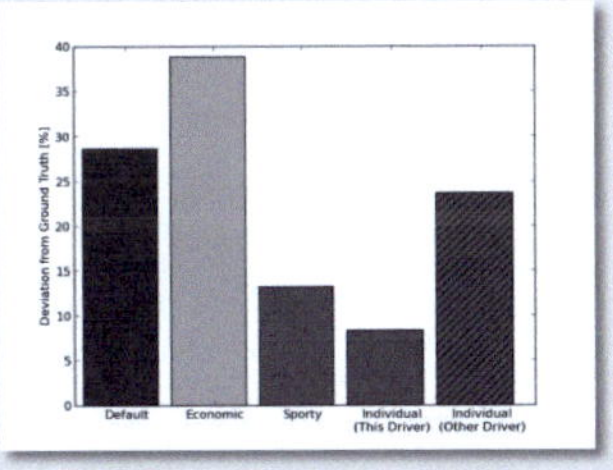	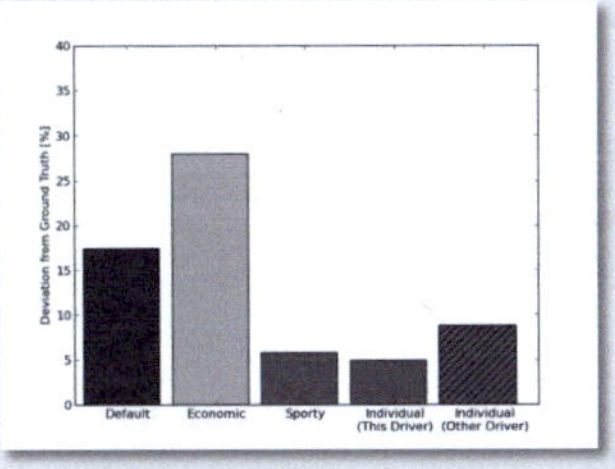

Table 2: *Comparison of two different drivers with their individual energy consumption estimation on the Turmberg Drive (~25 km). Note, that the distance travelled is measured by means of the accumulation of the GPS-Sensor signals, causing drifts in the travelling distance.*

Though the individual forecast performs best in all four cases, especially the drives on the Turmberg track showed, that influences as stop and go traffic, or busy highways, or the elevation profile of the track do have a high impact on the energy consumption estimation error, and should not be neglected for that reason. However, with consumption forecast could be improved by adjusting the consumption model to the individual driver.

4. Conclusion

In this work, the energy consumption of a given track is estimated with respect to the individual driving behavior of the driver. The driving behavior is, in this work, expressed as the individual velocity a driver chooses given a particular velocity restriction and the individual acceleration.
To forecast the energy consumption, this work prognosticates a virtual speed profile on the given track and calculates the energy consumption of the car on the virtual speed profile. The focus of this work is to enhance the speed profile calculation by the driver's habitual velocities given a particular speed limit and the driver's individual acceleration. Because the habitual parameters improve the pre-calculated speed profile during the drive, the energy consumption estimation improves during the drive, too (figure 4, left).
Though the influences on the energy consumption caused by the traffic situation and the elevation profile are more substantial, the energy consumption forecast could be improved by the individual adaption to the driver.

Acknowledgement
This work has been partially supported by the German Ministry of Science and Education (BMBF) in the project GeenNavigation under grant 13N12365.

References

[1] Tobias Bär, Ralf Kohlhaas, J. Marius Zöllner, and Kay-Ulrich Scholl. Anticipatory driving assistance for energy efficient driving. In 2011 IEEE Forum on Integrated and Sustainable Transportation Systems, pages 1–6, Vienna, 2011. Intelligent Systems and Production Engineering (ISPE), FZI Forschungszentrum Informatik, Haid-und-Neu Str. 10-14, 76131 Karlsruhe, Germany, IEEE.

[2] Tobias Bär, Dennis Nienhüser, Ralf Kohlhaas, and J. Marius Zöllner. Probabilistic driving style determination by means of a situation based analysis of the vehicle data. In 2011 14th International IEEE Conference on Intelligent Transportation Systems (ITSC), pages 1698–1703. IEEE, October 2011.

[3] I. M. Berry, Massachusetts Institute of Technology. Dept. of Mechanical Engineering, Massachusetts Institute of Technology. Technology, and Policy Program. The Effects of Driving Style and Vehicle Performance on the Real-world Fuel Consumption of U.S. Light-duty Vehicles. Massachusetts Institute of Tecnology, Department of Mechanical Engineering; and, (S.M. in Technology and Policy)–Massachusetts Institute of Technology, Engineering Systems Division, Technology and Policy Program, 2010.

[4] Bundesregierung Deutschland. Mobilität aus der Steckdose, 2011.

[5] Ralph Kampwirth. Jeder siebte Deutsche will 2020 ein Elektroauto fahren, 2012.

[6] Ralf Kohlhaas, Thomas Schamm, Dennis Nienhüser, and J. Marius Zöllner. Anticipatory energy saving assistant for approaching slower vehicles. In 2011 14th International IEEE Conference on Intelligent Transportation Systems (ITSC), pages 1966–1971. IEEE, October 2011.

[7] J. Van Mierlo, G Maggetto, E Van De Burgwal, and R Gense. Driving Style and Traffic Measures — Influence on Vehicle Emissions and Fuel Consumption. PhD thesis, 2004.

[8] M. Symmons, N. Haworth. Driving to Reduce Fuel Consumption and Improve Road Safety. In Road Safety Research, Policing and Education Conference, 2001, Melbourne, Victoria, Australia, page 7, 2001.

[9] Dennis Nienhüser, Tobias Bär, Ralf Kohlhaas, Thomas Schamm, Jochen Zimmermann, Thomas Gumpp, Marcus Strand, Oliver Bringmann, and J Marius Zöllner. Energy Efficient Driving and Operation Strategies Based on Situation Awareness and Reasoning. It - Information Technology, 54(1):5–16, 2012.

[10] Morgan Quigley, Ken Conley, Brian Gerkey, Josh Faust, Tully Foote, Jeremy Leibs, Rob Wheeler, and Andrew Ng. ROS: an open-source Robot Operating System, 2009.

8 Using Online Data from Local Traffic Management Centers to Assist an Energy-Efficient Driving Behavior

Dipl.-Ing. **Mario Krumnow**, Dipl.-Ing. **Robert Richter**, Dipl.-Ing. **Per Lewerenz**, Dipl.-Ing. **Torsten Schubert** und Dipl.-Ing. **Stephan Uebel**, Technische Universität Dresden, Dresden, Germany

Abstract

The main goals of today's research and development are leading to different systems and topics for more energy-efficient technologies in powertrains and intelligent driver assistance systems. The funded project "Energieeffizientes Fahren 2014" (EFA 2014/2) aims for an increasing range of electric vehicles. In order to reach this goal an approach has been chosen which includes infrastructure data using Vehicle-to-Infrastructure (V2I) communication technologies. Particularly traffic actuated traffic lights are being utilized since these are state of the art to optimize traffic flows. Based on the interaction between vehicle and infrastructure the driver will be able to adapt an energy-efficient manner of driving easily through additional information and integrated board aggregation. This approach has been successfully tested in Dresden.

Kurzfassung:

Im Fokus derzeitiger Forschungen im Bereich der Elektromobilität steht die Reichweitenerhöhung von E-Fahrzeugen, dabei sollen intelligente Fahrerassistenzfunktionen eine zu einer energieeffiziente Fahrweise hinführen. Das Förderprojekt „Energieeffizientes Fahren 2014" (EFA 2014/2) verfolgt in diesem Zusammenhang den Ansatz, die Energieeffizienz und somit die Reichweite von Elektrofahrzeugen durch Einsatz von modernen V2I-Technologien zu erhöhen. Dabei werden Daten der Umgebung (Lichtsignalanlagen, Verkehrslage) an das Fahrzeug übermittelt, welches so eine geeignete Fahrstrategie wählen kann. Als Verkehrsbeeinflussungsanlagen im Stadtverkehr werden zunehmend verkehrsabhängige Lichtsignalanlagen (LSA) eingesetzt, welche maßgeblich den Verkehrsfluss bestimmen. Aufgrund dessen werden insbesondere Schaltdaten dieser Form von Lichtsignalanlagen im Rahmen von EFA 2014/2 herangezogen und an das Fahrzeug übertragen, um so den Fahrer zu unterstützen und seine Fahrweise anzupassen. Auf einer Pilotstrecke in Dresden von mehreren Kilometern Länge konnte dieser Ansatz erfolgreich getestet werden.

Keywords: *Traffic Management, Communication, GPS, UMTS, DATEX II, Energy efficient, Microscopic Traffic Simulation, Dresden*

1. Motivation

Traffic demand is increasing constantly and with it emission density which leads to more and more environmental pollution. In order to reduce arising emissions due to road traffic electric vehicles are being introduced. The disadvantage of these vehicles is the relatively short range. The Research and Development (R&D) project EFA 2014/2 aims to increase the range by adapting energy-efficient driving behaviors e.g. precise speed recommendation while approaching a traffic light. Therefor an advanced driver assistance application has been developed. It comprises the interaction between a vehicle and its surroundings – especially traffic lights – based on Vehicle-to-Infrastructure (V2I) communication since speed recommendations are only possible by knowing upcoming signal states of the traffic light ahead.

2. Online infrastructure data sources

Traffic management actions require many information that can be obtained e.g. from induction loops, cameras and floating car data (FCD). Utilizing this detected data infrastructure components as traffic lights or variable message signs can be influenced and adapted in order to affect the current traffic flow. Since energy consumption is dependent on traffic flow which is mainly controlled by traffic lights in urban areas, it is necessary to understand how traffic light controls work. A new approach is being established which aims to change the behavior of a single vehicle instead of the traffic flow. Therefor the ascertainment of an individual energy-efficient driving behavior for every single vehicle is being pursued depending on traffic signal states. Signal times and delay times as the most important characteristics are being presented in the following chapters.

2.1 Prediction of signal states

In order to determine an energy-efficient driving behavior while approaching a traffic light, it is essential to know the signal state at the time the vehicle arrives at the stop line. This implicates the necessity of predicting future signal times and not current ones. Since most of the existing traffic lights are traffic actuated, signal times are not fixed and the use of probabilities is required [6]. Traffic actuated traffic light controls contain different signal programs depending on traffic demand which varies from morning over midday to evening hours (figure 1).

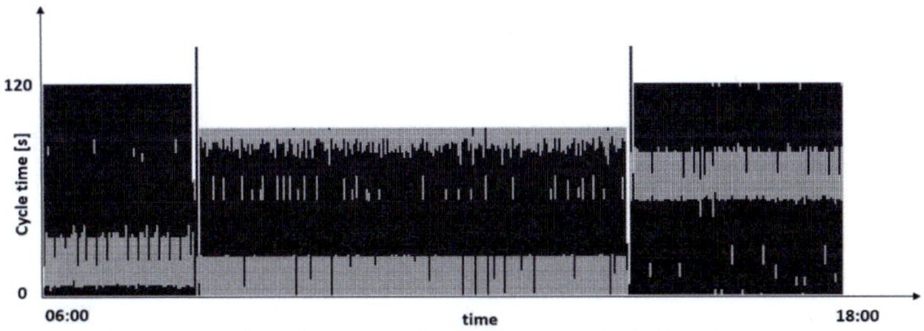

Figure 1: *Signal states of a traffic actuated signal program*

Within these signal programs there is also the possibility of requesting additional phases by public transport or little frequented directions causing shifted signal times.

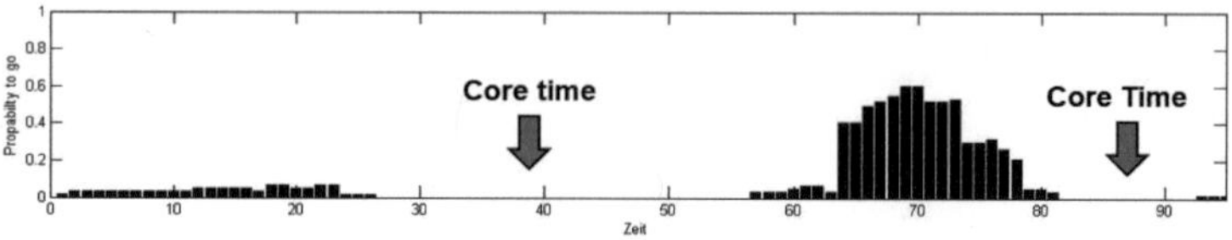

Figure 2: *Example of "probability-to-go" values*

To ascertain the predicted probabilities to go, signal states are being simplified to only two states and coded in form of binary vectors. Therefor, "0" relates to the states red, amber and red-amber, and "1" refers only to the state green [7]. This vector composed over a time period of several minutes is being transmitted to the vehicle where energy-efficient driving behavior is being computed (figure 2).

2.2 Delays due to traffic light signals

Another important fact is the knowledge of congestion in front of traffic light controlled intersections. There are little possibilities measuring the queue length, e.g. with cameras, so that is it more common to estimate the values of queue lengths. The used estimation procedure depends on the disposability of data sources. Statistic and heuristic methods are being applied in case only historic or aggregated data is available. If data of fixed detectors e.g. induction loops are available, approaches like the method of Mück can be utilized [10].
If mobile sensors as FCD are the basis for the prediction, the method of Neumann can be applied [11]. Within this funded project static sensors are being used in order to estimate delay time in seconds and queue length in meters. Both values are being updated every second. In addition to signal times, also congestion affects the traffic light control so that the knowledge of queue length and delay time is essential.

3. Communication Chain and Car Positioning

This part describes the fundamental and technical challenges within the communication chain from traffic light into the vehicle. This includes especially the latency of required information between the traffic lights and the car. Furthermore the directly conjunction of the vehicle position and the needed information in urban scenarios is also a challenge.

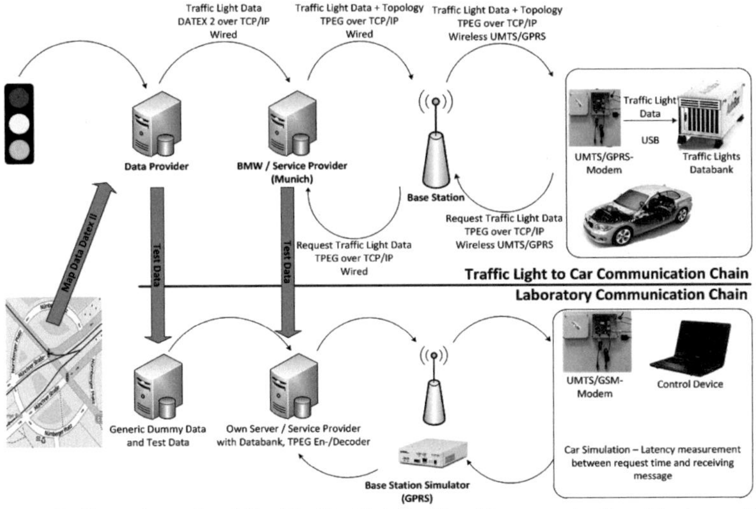

Figure 3: *Overview about the Traffic Light to Car Communication Chain and Laboratory Communication Chain*

These aspects can be well explained by the following example of a driving situation at a multitrack intersection in an urban scenario. The vehicle positioning solution based on GNSS is very poor causing by strong multipath urban environment. Due to this, the positioning solution doesn´t allow a relation of the real driven track. In this case, the car computer doesn´t know which traffic light information (e.g. red or green remaining time) must be displayed for the driver. So, the driver gets all possible traffic light information and must choose the right one. For avoiding this, a track selection is needed. Furthermore, the displayed traffic light has to be the right one. For this, the whole technical system has to handle the delayed information over the complete communication chain between traffic light and car.
For the communication different types of systems (e.g. GSM/UMTS) and protocols (e.g. TPEG TSI - Transport Protocol Experts Group Traffic Service Information) are used. It´s a centralized approach, who defines a central service provider which handles and distributes the needed data. Figure 3 gives an overview on the complete communication chain between traffic light and car. A complete reproduction of the communication chain in a laboratory allows simulation and measurements of the latency between different parts of the chain. So, testing of different latency time measurements and interference scenarios and their solutions for communication are possible.
For car positioning two approaches, Kalman- and Particel Filter, have been investigated. These two filters are state of the art [18] and were combined with an enhanced digital map and preprocessed video data [4].

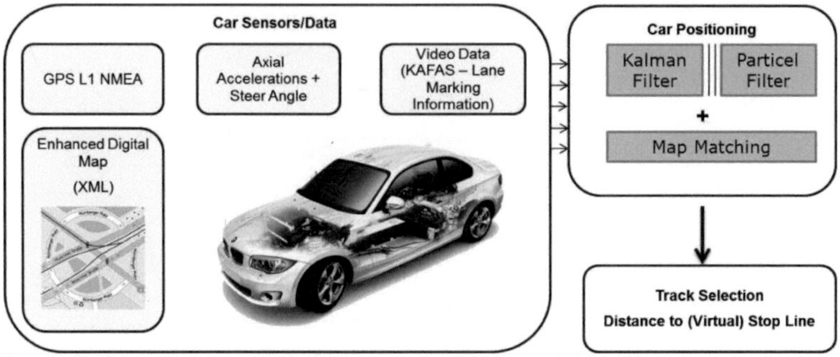

Figure 4: *Overview about the car sensors/data and the using positioning algorithms for track selection and distances to (virtual) stop lines*

The video data contains information about the distances to the left and the right lane marks. Furthermore the enhanced digital map provides all lanes for the test scenario in Dresden with additional information like stop markings. This information is deposited as an XML scheme and can be easily virtualized, combined and overlain with other maps. Figure 4 gives an overview about the used car sensors and sensor data. The data is provided via CAN bus to the car computer and is synchronically sampled for the filters. The idea is to run two filters in parallel for using estimated solutions for track selection. Combined with the additional video data, different hypotheses for lane selection will be examined based on the enhanced digital map. In order to assess the performance of the methods, synthetic data for filter calibration is used at first. After that, real data from the test scenario environment is used for validation. Due to this, a submeter precision ($\leq$0.5 m for $\pm 2\sigma$) can be achieved for car positioning in the case of the test environment in Dresden. The value 2σ means the probability of the positioning values is better or equal than 0.5 m in 95.4% of cases. In 4.6 % of cases the probability of the positioning values is worse than 0.5 m.

4. Human Machine Interface

This chapter deals with the way how the above shown traffic information can be displayed to the driver to influence him to an optimal driving behavior.

4.1 Experimental Vehicle

To validate the developed measures in real traffic an experimental vehicle is used. It is a full electric car manufactured by BMW called ActiveE. A picture of the vehicle can be seen in figure 5. It is a converted BMW 1 Series Coupe (E82). The electric motor drives the rear axle with up to 125 kW. The vehicle has a lithium-ion-battery with 32 kWh for an electric range up to 160 km.

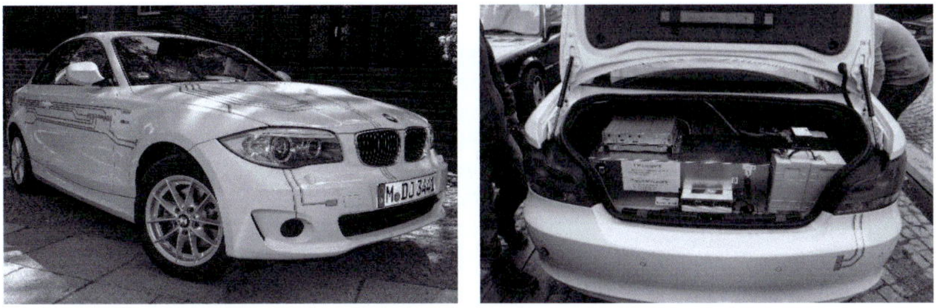

Figure 5: *Experimental Vehicle with in-vehicle measurement system.*

It is modeled in MATLAB/SIMULINK, to have the opportunity to simulate the energy consumption of the vehicle at different traffic conditions. Additional information on the model can be found in [15].

4.2 Human interaction

By the merging of vehicle internal and infrastructure information in future driver assistance functions the amount of information for drivers will increase significantly, as the present research project shows.

It is important to make sure that it does not lead to excessive demands of the driver by too much information. Furthermore, it should be noted that the information, will not always be one hundred percent reliable due to the source. Therefore, the forms of representation must be chosen so that the driver asses the information as helpful even if they are uncertain. This means that the form of representation has a significant impact on the overall acceptance of the system. In the present paper, the approaches are shown to communicate uncertain information for the efficient driving in traffic light approach situations to the driver. As information channels used both for the visual display in the dash panel as well as the haptic feedback via an active accelerator pedal. The installed dash panel of the test vehicle used can be programmed to examine flexible different forms of representation. Near-series illustrations are presented to the driver and the impact on driver behavior through prototypical auxiliary displays can be avoided. With the active accelerator pedal the counter pressure, which the driver has to apply to the operation of the pedal, can be varied dynamically. On the one hand, a direct influence of the driver can be initiated; on the other hand the driver's attention can be stimulated by vibrating the accelerator pedal. For the optical information two areas were defined in the display. One area is for the recommendation, and the other for information (figure 6).

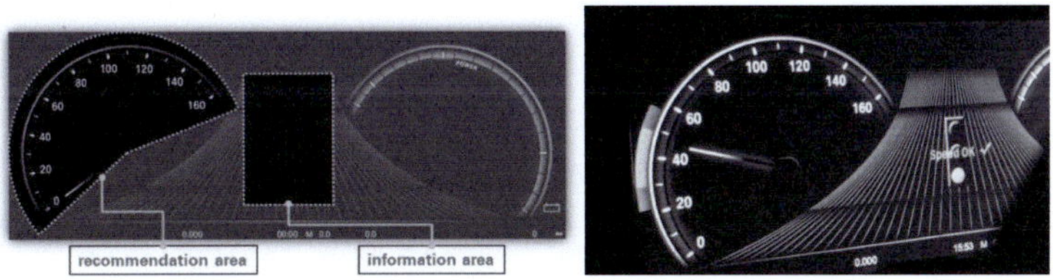

Figure 6: *Programmable digital instrument panel*

In the area of recommendation to a range of traveling speed is recommended. With the recommended speed, the next traffic light is reached at a green phase without vehicle standstill. The quality of information is taken such a form into account that a distinction is made between the core times (prob2go > 90%) and a region of high probability (90% > prob2go > 70%). The representation in the information area depends on the current speed relative to the recommended speed. Therefore, the consideration of information quality in the information area is set automatically. If the speed of the vehicle is below a certain level (10km/h), the remaining phase duration is displayed. The assignment and the information displayed are shown in figure 7.

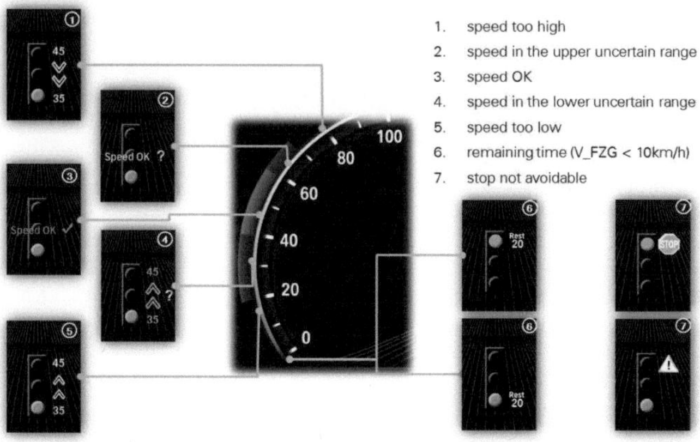

***Figure 7:** Assignment of information and recommendation*

Matching the information of the different areas, the active accelerator pedal is triggered. In cases 1) and 5) the driver is informed by a slight vibration of the pedal that he should optimize its longitudinal dynamic behavior. In cases 2) and 4) the driver is caused by the increase or decrease of the pedal counter pressure to change the speed to the optimal range.

4.3 Validate the benefit of driver assistance in real traffic scenarios

It is very difficult to get reproducible results while validating the effect of driver assistance systems in real traffic. The simulation is a very powerful way to analyze these systems.

Current research in [2], [13] and [16] in the context of traffic light assistance systems (TLAS) und predictive cruise control systems [1] show the high potential of driver assistance systems to realize an energy efficient driving behavior. All of these systems use the information of traffic lights to calculate an optimal velocity to approach the intersection. Consumption minimizations between 3 and 5% [2] and in some situations about 30% [16] in comparison to the uninformed driver are shown.It is essential to examine the influence of other road users in particular the queue length at the stop line of traffic lights. Currently the scenarios examined are only very simple traffic situations (e.g. one lane, static traffic light programs).

To simulate more complex traffic scenarios a new simulation framework [14] is developed. It is an interface between detailed nanoscopic vehicle simulation with MATLAB/Simulink and traffic flow simulation with SUMO. With different parameters like speed limits, traffic light control, number of lanes, and density of traffic flow the impact of different traffic situations on the energy consumption can be determined. For that purpose a vehicle model (see chapter 5) is used to determine the energy. Different traffic scenarios and there influence on the individual energy consumption are analyzed. The results will show whether these systems are useful and how much the benefit is.

5. First Results

In a first realistic use case the potential of TLAS is shown. A part of an urban route in Dresden with a constant traffic stream is modelled in the SUMO Simulation suite. On basis of [1] a traffic light assistance system is implemented in MATLAB and simulated with the described framework in chapter 4.3. Different parameters like the time of occurrence of the analyzed vehicle are varied.

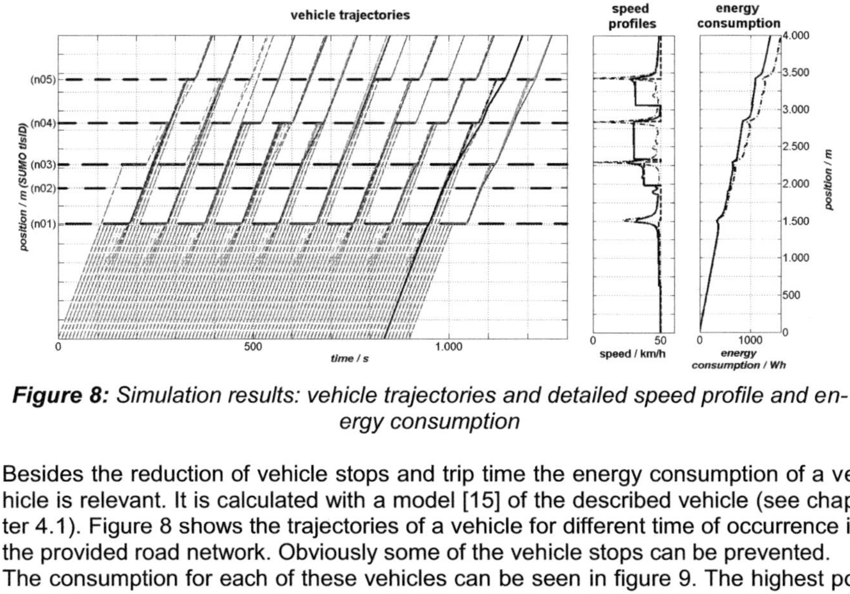

Figure 8: *Simulation results: vehicle trajectories and detailed speed profile and energy consumption*

Besides the reduction of vehicle stops and trip time the energy consumption of a vehicle is relevant. It is calculated with a model [15] of the described vehicle (see chapter 4.1). Figure 8 shows the trajectories of a vehicle for different time of occurrence in the provided road network. Obviously some of the vehicle stops can be prevented.
The consumption for each of these vehicles can be seen in figure 9. The highest potential for energy consumption is provided for vehicle number 94. Although there are obviously some uncomfortable situations that could cause higher energy consumption (see vehicles 32, 56 and 57 in figure 9).

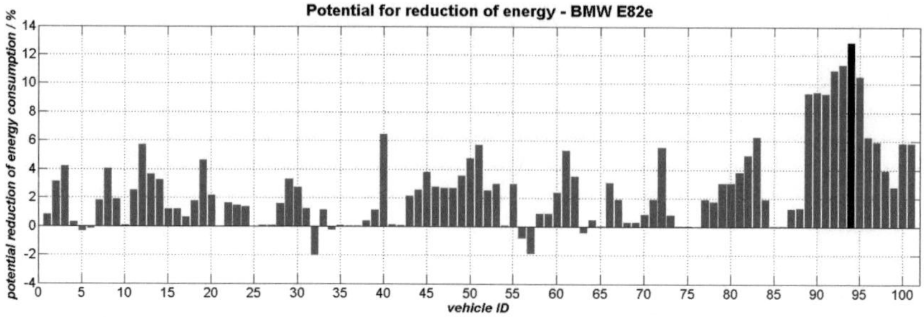

Figure 9: *Simulation results: Energy consumption of simulated traffic situations. Vehicle 94 is highlighted with the best potential*

These are caused by the effect that in some cases a vehicle with TLAS could pass the next traffic light but this results in an uncomfortable situation at the next traffic light, so the TLAS cannot avoid the vehicle to stop. This demonstrates the need of further research on that topic.

6. Inner Vehicle Measures

The above shown measures are sufficient to apply an energy-efficient driving behavior for electric or conventional vehicles. Taking modern hybrid cars into consideration an optimal control strategy has to be applied additionally.

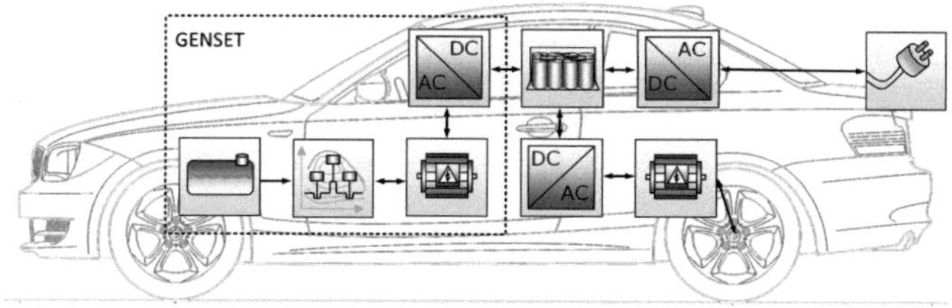

Figure 10: *Powertrain of the EREV model*

To examine control strategies for hybrid cars the model of the experimental vehicle introduced in chapter 4.1 is changed to an extended-range electric vehicle (EREV) by adding a generator set (GENSET). This set consists of engine and an electric generator with power inverter. The structure of the powertrain is shown in figure 10.
A main focus lies on the combustion engine of the generator set. The cold starting behavior has a great impact on emissions that cannot be neglected by the control strategy. So the analysis of a suitable combustion engine is necessary. A conversion of a 4-cylinder '1.4 TSI' engine (EA211) of Volkswagen is used on a test bench. The converted engine is a 3-cylinder aspirated direct injection gasoline engine with a displacement of 1 liter to fit the engine power to the requirements [9], [3]. The results of the test bench are needed to create a model in GT-SUITE which is able to develop

engine-specific control strategies und complete the data base for a powertrain control strategy.
The developed control strategy is based on the Equivalent Consumption Minimization Strategy (ECMS) that was introduced in 2000 by Paganelli [12]. This strategy is originally meant for a parallel hybrid electric vehicle with a small battery. The key idea is to find an equivalence between the energy stored in the battery E_{Bat} and the energy stored in the fuel tank E_{Fuel}. The result is an equivalent energy

$$E_{\lambda} = E_{Fuel} + \lambda \cdot E_{bat}$$

where λ is the equivalence factor. The power of the GENSET, which can be chosen freely, has to be determined so that E_{λ} is minimal. The ECMS can be derived from Pontryagin's Maximum Principle, which is a method of the calculus of variations [17].

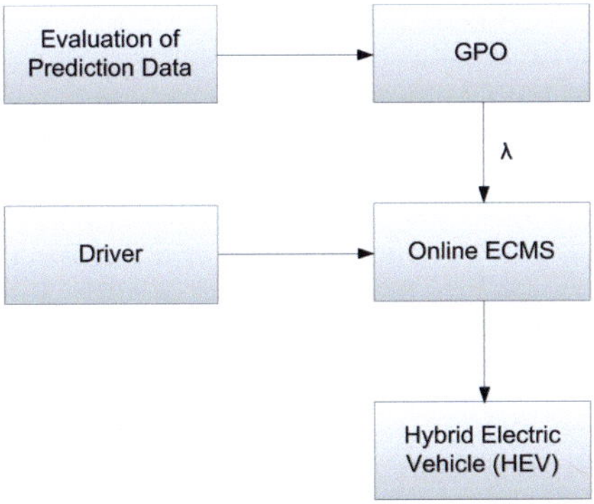

Figure 11: *Structure of Control Strategy*

An issue is to find the equivalence factor that is different for every track (figure 11). Therefor an iterative approach called Global Pontryagin Optimization (GPO) is used that calculates λ out of prediction data as seen [17], [8].

7. Conclusions

The R&D project EFA 2014/2 is pursuing the approach to obtain infrastructure information of the whole road network, e.g. traffic light information. This information are being broadcasted using the widespread and available mobile communications network. This involves assets and drawbacks e.g. concerning the availability and higher latency times compared to usual directional Wi-Fi connections. In future, needed data streams can be distributed over different communication systems. Therefore, none time critical information (e.g. map data) can be distributed by broadcasting systems like DAB+. For time critical Information (e.g. short-term prediction data) it seems to be suitable to use communication systems with low latency like LTE. In urban areas traffic actuated traffic lights are being utilized now and in the future even more. These traffic light controls are being influenced by traffic flows as well as single vehicles so that it is possible to react dynamically to varying traffic demands. Due to this variation, only a prediction approach is capable. In doing so it is necessary to create an

approach that is able to deal with probability-based values. Also the developed energy-efficient driving strategy needs to be adapted to these new conditions.
Results of various simulations show an energy saving potential of about 10 percent by using this developed approach. This kind of driving strategy can also lead to an increasing energy consumption in case the provided data is uncertain. Nevertheless, it could be shown that a wide transmission of signal data can lead to a global optimum regarding to the driving strategy for the whole route. The state of the art is to optimize driving strategies while approaching single traffic lights. Considering all traffic lights on a route may lead to even higher saving potentials. In order to achieve the desired driving strategy the driver is being supported actively by the system. On the one hand, the driver gets driving instructions via a display which can be adopted; on the other hand, an on-board unit pays attention to an efficient realization of these instructions. Therefore, a special display has been developed that is able to deal with probability-based values.
Finally an energy-efficient approach has been created to increase the range of electric vehicles which is transferable to other cities. But what is going to happen if infrastructure as well as vehicles get smarter at the same time?

Bibliography

[1] Asadi, B.; Vahidi, A. (2010). Predictive Cruise Control: Utilizing Upcoming Traffic Signal Information for Improved Fuel Economy and Reducing Trip Time, Vol. 19. In: IEEE Transactions and Control Systems Technology, S. 707–714.

[2] Bley, O.; Kutzner, R.; Friedrich, B. (2011). Kooperative Optimierung von Lichtsignalsteuerung und Fahrzeugführung. In: AAET 2011 Automatisierungs-, Assistenzsysteme und eingebettete Systeme für Transportmittel.

[3] Einert, D. (2014). Experimentelle Untersuchungen zum Kaltstart-Warmlaufverhalten an einem Range-Extender Versuchsträger, Technische Universität Dresden – Institut für Automobiltechnik Dresden – Lehrstuhl Verbrennungsmotoren, Dresden

[4] Gosda, U.; Weber, R.; Michler, O.; Zeisberg, S.; Mademann, E. (2013). Target tracking in wireless sensor networks by data fusion with video-based object detection, Positioning Navigation and Communication (WPNC), 2013 10th Workshop on, 20-21.

[5] Guhr, C. (2011). Verbesserung von Effizienz und Dynamik eines hubraumkleinen turboaufgeladenen 3-Zylinder-DI-Ottomotors durch Abgasrückführung und ein neues Ladeluftkühlkonzept, Technische Universität Dresden – Institut für Automobiltechnik Dresden – Lehrstuhl Verbrennungsmotoren, Dresden

[6] Krumnow, M (2012). Schaltzeitprognose verkehrsadaptiver Lichtsignalanlagen im Rahmen des Projektes EFA 2014/2, 8. VIMOS Tagung, Dresden

[7] Krumnow, M (2014). Schaltzeitprognose verkehrsabhängiger Lichtsignalanlagen im Rahmen des Forschungsprojektes EFA 2014/2, Tagungsbericht Heureka 2014, FGSV Verlag, Köln

[8] Kutter, S.; Bäker, B. (2011). An iterative algorithm for the global optimal predictive control of hybrid electric vehicles. In: Vehicle Power and Propulsion Conference (VPPC), IEEE

[9] Mandausch, M (2013). Aufbau und Inbetriebnahme eines für Range-Extender-Betrieb vorgesehenen Versuchsträgers, Technische Universität Dresden – Institut für Automobiltechnik Dresden – Lehrstuhl Verbrennungsmotoren, Dresden

[10] Mück, J. (2002). Schätzverfahren für den Verkehrszustand an Lichtsignalanlagen unter Verwendung halteliniennaher Detektoren, Tagungsbericht Heureka 2002, FGSV Verlag, Köln

[11] Neumann, T. (2011). Rückstaulängenschätzung an Lichtsignalanlagen mit Floating-Car-Daten, Berichte aus dem DLR-Institut für Verkehrssystemtechnik, Berlin

[12] Paganelli, G.; Guerra, T.; Delprat, S.; Santin, J., J.; Delhom, M.; Combes, E. (2000). Simulation and assessment of power control strategies for a parallel hybrid car, in: Proceedings of the Institution of Mechanical Engineers, Part D: Journal of Automobile Engineering, Mechanical Engineers, 214(7), 705–717

[13] Schubert, T. (2010). Entwurf und Evaluierung einer prädiktiven Fahrstrategie auf Basis von Ampel-Fahrzeug-Kommunikationsdaten. Diplomarbeit, Technische Universität Dresden.

[14] Schubert, T.; Krumnow, M.; Bäker, B.; Krimmling, J. (2013). Using Nanoscopic Simulations to validate the Benefit of Advanced Driver Assistance Systems in complex Traffic Scenarios. Proceedings of the 3rd International Conference on Models and Technologies for Intelligent Transportation Systems 2013, Verkehrstelematik. Vol. 3, TUDpress, Dresden, ISBN: 978-3-944331-34-8.

[15] Schubert, T.; Uebel, S.; Krumnow, M.; Bäker, B.; Krimmling, J. (2014). Analyse eines LSA-Assistenzsystems mittels nanoskopischer Simulation in komplexen Verkehrsszenarien, AAET - Automatisierungssysteme, Assistenzsysteme und eingebettete Systeme für Transportmittel, Braunschweig.

[16] Schuricht, P.; Michler, O.; Bäker, B. (2011). Efficiency-increasing driver assistance at signalized intersections using predictive traffic state estimation, In: 14th International IEEE Conference on Intelligent Transportation Systems (ITSC). DOI: 10.1109/ITSC.2011.6083111, S. 347 – 352.

[17] Uebel, S.; Tempelhahn, C.; Liebers, M.; Kutter, S.; Bäker, B. (2014). Anwendung der Variationsrechnung für Steuerungsaufgaben im Kraftfahrzeug, in: at–Automatisierungstechnik 2014, 62(4), 249-259

[18] Bar-Shalom, Y.; Li, X.; Kirubarajan, T. (2001). Estimation with Applications to Tracking and Navigation: Theory Algorithms and Software.

9 Improvements on the Accuracy of Electro-mechanical Simulation of Series-parallel Hybrid System Power Train - Detailing the motor system including simulation model for the inverter and boost converter

M.eng. **W.H. Yang**, B.eng. **H. Sato**, Prof. PhD.eng **Y. Kamiya**,
Prof. **Y. Daisho**, WASEDA University, Tokyo, Japan

Abstract

This paper addresses the improvements on the accuracy of the simulation model by detailing the motor system including the inverter and the boost converter of Toyota Prius PHV(2012), one of most effective and popular Series-parallel type plug-in hybrid vehicles. It devises a method that can calculate the inverter and boost converter losses with high accuracy by estimating the output current and regenerative current of the motor system. In the past, we have used the actual measurement data such as efficiency map to calculate the loss of a motor system. However, this method allows us to separate the one loss into three different categories, i.e. the losses caused by the inverter, the boost converter and the motor. Based on the separation of the motor system losses, we can then build a multi-voltage motor system or estimate the losses of the different ranks of vehicles like buses or trucks.

1. Introduction

In recent years, due to the awareness of environmental protection and the on-going energy resource deficiency, plug-in hybrid vehicles (PHVs) have drawn world attention as it is hoped for to relive the matter. Comparing to other types of power trains, plug-in hybrid vehicles are developed at a relatively faster pace as it has longer cruising distance and lower power recharging facility problems, which increases its usability making it more convenient for users and hence the willingness to adopt such technology. Although hybrid powered train technologies did not converge towards a single solution, the series-parallel type hybrid vehicle has proved successful in fuel efficiency and automobile market.

Previous research constructed series-parallel method (SP method) and series method (S method) hybrid system simulators for existing vehicle models using MATLAB/Simulink [1]. Other research then used the developed simulators to optimize the installed components and control methods of hybrid systems with the same base vehicles [2]. Subsequently, considering possible long-term trends of PHV technology development, it has become necessary to expand the scope of research to include commercial and range-extender type vehicles. However, the developed simulators are difficult to apply to these types of vehicles in their current configurations.

Therefore, this paper describes the results of a study related to more detailed modeling of the motor system in a SP method PHV. In addition to the creation of a more detailed simulation model, this development also aimed to construct a highly versatile simulator that can be applied to different vehicle categories without depending on actually measured maps and the like [3].

2. Research Method

As shown in Fig. 1, previous research calculated motor system loss based on actually measured maps. In contrast, the research described in the following sections aimed to calculate individual losses from the motor, inverter, boost converter, and other components. The purpose of these individual calculations is to help expand application of the simulator to other vehicle categories, multi-voltage systems, and the like. This aim was accomplished by adopting a new motor current estimation method, and by constructing new simulation models for the inverter, boost converter, and other components. Fig. 2 shows the equivalent circuit of the studied motor system [4].

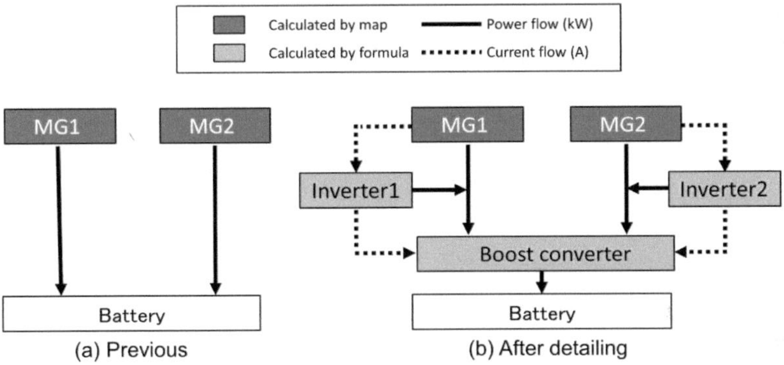

Figure 1: *Simulation model*

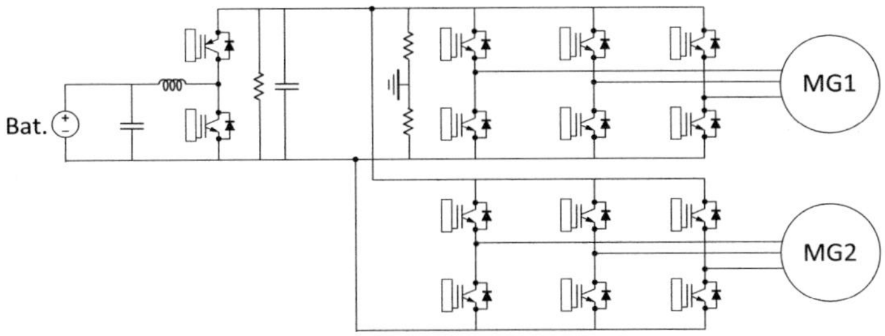

Figure 2: *Motor system circuit*

3. Research Results

3.1 Estimation of Motor System Current

First, this research aimed to estimate the internal current of an interior permanent magnet synchronous motor (IPMSM), a type of motor that is frequently used in modern PHVs and BEVs.

Generally, IPMSMs use a current vector method for control (also called the d-q axis control). In this research, this method was assumed to incorporate both maximum torque per ampere control (MTPA) and field weakening control [5].

In basic terms, input current is determined in accordance with the MTPA line as shown in Fig. 3 (the A-O line). However, field weakening control must also be performed to increase the motor speed. In this case, the motor speed can be increased by increasing the d-axis current to weaken the field of the permanent magnet (A to D, B to B', or C to C'). At the same time, the input current is limited by the inverter (the current limit circle and voltage limit ellipse).

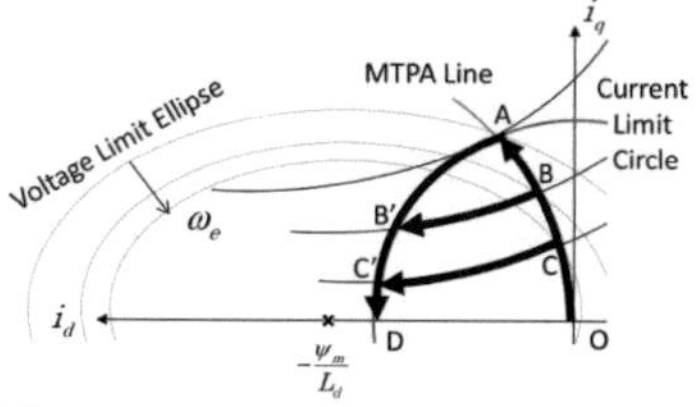

Figure 3: *MTPA & Field weakening control*

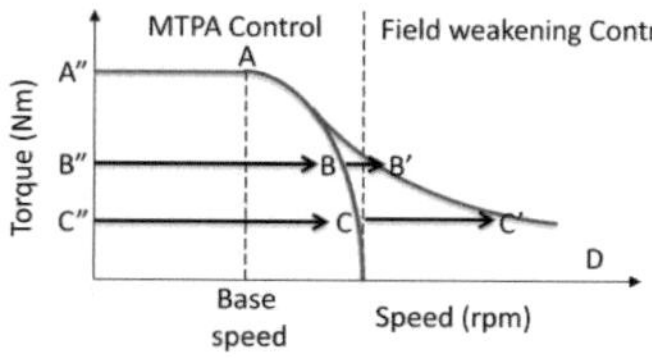

Figure 4: *Method to estimate the motor system current*

To estimate the motor system current with this control method, first, the input current of the MTPA control is assumed using data from locked rotor tests. The input current values A", B", and C" in Fig. 4 is corresponding to the torque in the locked rotor tests. Since values A, B, and C operate along the same MTPA line as A", B", and C", these form the same currents as A", B", and C". Next, since A, B', and C' are at the current limits, the same current values are formed. Finally, the current between B and B', and C and C' are corrected and calculated by linear interpolation. Fig. 5 shows the input current estimated by this method.

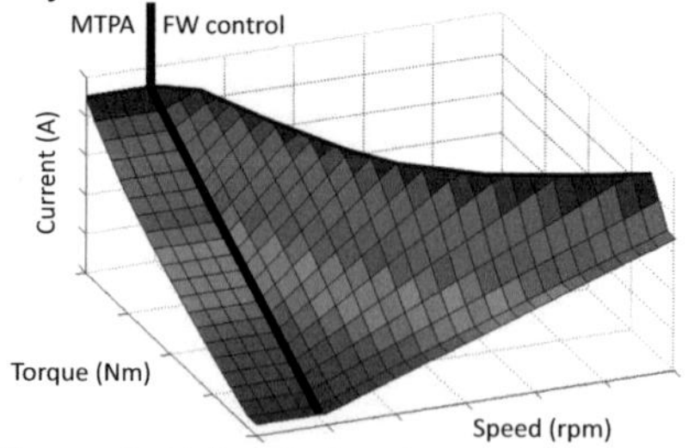

Figure 5: *Estimated motor system current*

For verification, the estimated input current was compared with actually measured motor input current using a commercial SP method hybrid vehicle (HV) [6]. First, the results in Fig. 6 confirm that the current obtained with respect to torque in the locked rotor tests matches the input current in the MTPA control region. In addition, the results in Fig. 7 show that the upper limit speed of MTPA control is approximately twice that of the base speed. This shows that the motor operating point in the JC08 test cycle can be controlled almost entirely by MTPA control.

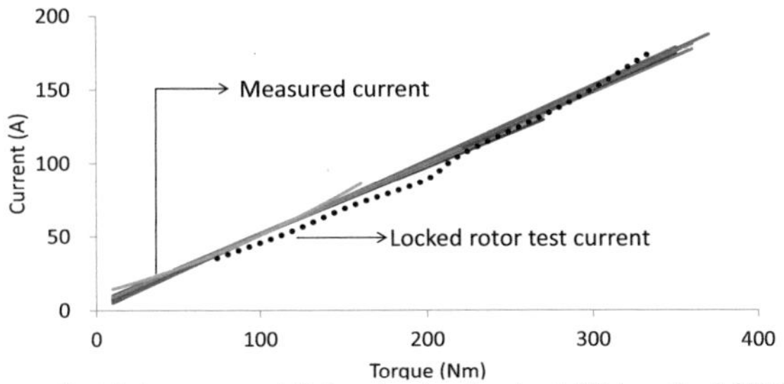

Figure 6: *Motor current & Locked rotor test (Prius 2nd (2004))*

In contrast, the estimated current during field weakening control does not match the results of past research. According to the current state of field weakening control in Fig. 7, the current on the max operating curve is not up to the value of the current limit circle, indicating that it may have been applied another control method such as maximum torque per voltage control (MTPV)(Fig. 8), which allows motor operate in higher speed area whilst lowering the iron loss. If the MTPV control is applied, the center of the voltage limit ellipse must be inside rather than outside the current limit circle assumed at first. Although the control method in field weakening is known, it is still difficult to estimate the current value by simple methods such as MTPA control. However, since motor control in the JC08 test cycle can be performed almost entirely by MTPA control, this has little effect on the results of the simulation.

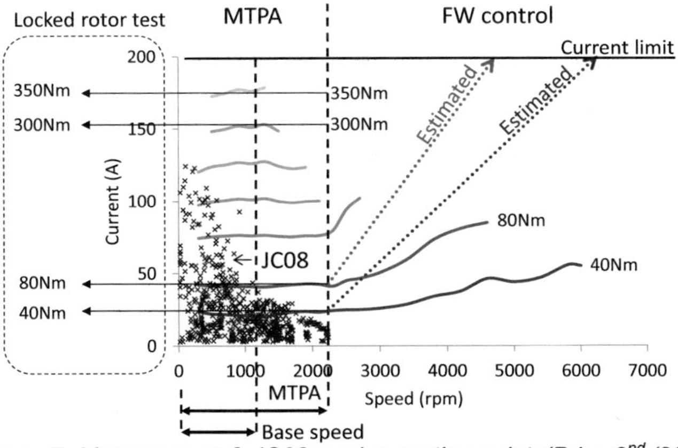

Figure 7: *Motor current & JC08 mod operating points(Prius 2nd (2004))*

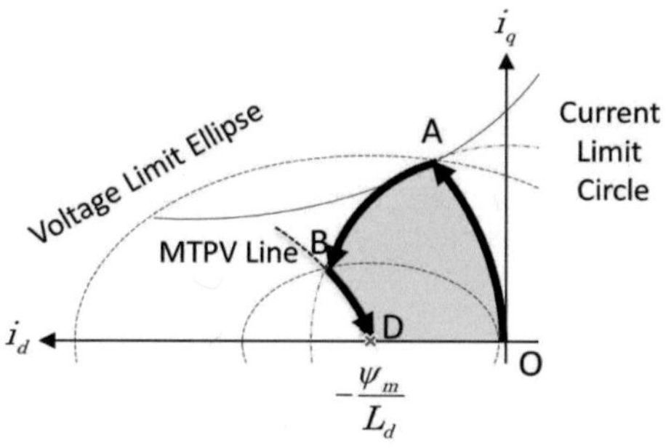

Figure 8: *MTPV control*

Based on the analysis above, the motor input current was estimated as shown in Fig. 9 in accordance with the following approach.

1. The values obtained through locked rotor tests were used to express the relationship between torque and input current in the MTPA control.
2. The upper limit speed of MTPA control was set to approximately twice the base speed.

The results of this estimation showed that motor operation in the JC08 test cycle is almost entirely within the MTPA region.

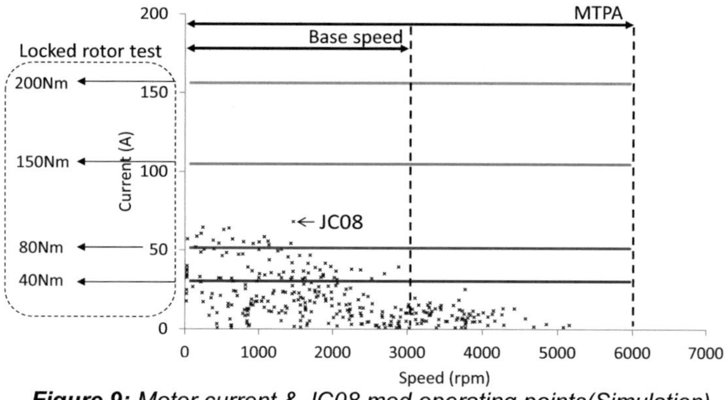

Figure 9: *Motor current & JC08 mod operating points(Simulation)*

3.2 Inverter Loss Modeling

Fig. 10 shows the equivalent circuit of the studied inverter. The inverter contains six insulated gate bipolar transistor (IGBT) modules and three smoothing capacitors. The loss during inverter operation is calculated to obtain the efficiency with respect to the load current. Specifically, this refers to the loss of the IGBT modules and the loss due to the dc resistance of the smoothing capacitors.

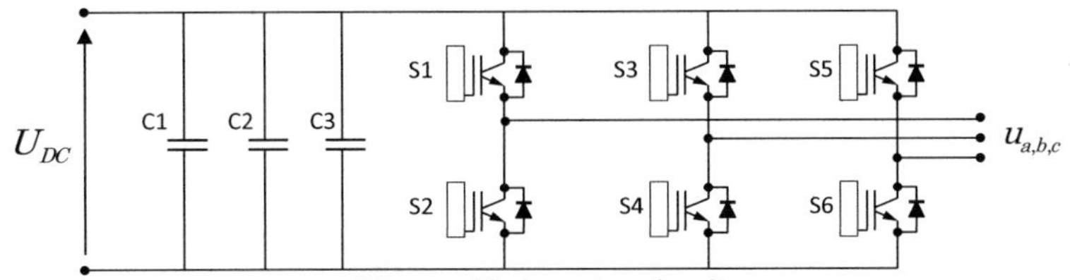

Figure 10: *Inverter circuit*

The loss of the IGBT modules was assumed to be the total loss of the IGBTs and the freewheeling diodes (FWDs). There are three types of IGBT loss in two separate categories: steady state loss and switching loss (comprising both turn-on loss and turn-off loss). Each type of loss is calculated as shown in (1) to (4) [7].

$$P_{con}^{IGBT} = i_{out} \cdot u_{sat}^{IGBT} \cdot \left(\frac{1}{8} + \frac{m}{3\pi} \cos\varphi \right) \tag{1}$$

$$P_{SW}^{IGBT} = \frac{f_s}{2} (E_{on} + E_{off}) \tag{2}$$

$$P_{con}^{Diode} = i_{out} \cdot u_{sat}^{Diode} \cdot \left(\frac{1}{8} - \frac{m}{3\pi} \cos\varphi \right) \tag{3}$$

$$P_{re}^{Diode} = \frac{f_s}{2} E_{re} \tag{4}$$

where,

i_{out}: Output current (A)
u_{sat}^{IGBT}: IGBT saturation voltage (Volt)
m: Modulation index
$\cos\varphi$: Power factor
f_s: Switching frequency (Hz)
E_{on}: Turn-on loss (J/pulse)
E_{off}: Turn-on loss (J/pulse)
u_{sat}^{Diode}: Diode forward voltage (V)
E_{re}: Recovery loss (J/pulse)

Equation (5) shows the calculationformula for smoothing capacitor loss. The pulsating current (iippler) is calculated at approximately 50% of the output current. The equivalent series resistance (ESR) is the equivalent dc resistance.

$$L_C = i_{ippler}^2 \cdot ESR = \left(0.5033 \cdot i_{out}\right)^2 \cdot ESR \tag{5}$$

The 6MBI300V-120-50 module fabricated by Fuji Ceramics Corporation was used to set the IGBT module parameters [8]. The dc resistance of one smoothing capacitor was set to 10 micro-ohms, the modulation index and power factor were set to 1, and the switching frequency was set to 5 kilo-hertz.

Fig. 11 compares the simulation results with actually measured data [4]. Although the overall estimated efficiency is on the high side, the trends closely match each other.

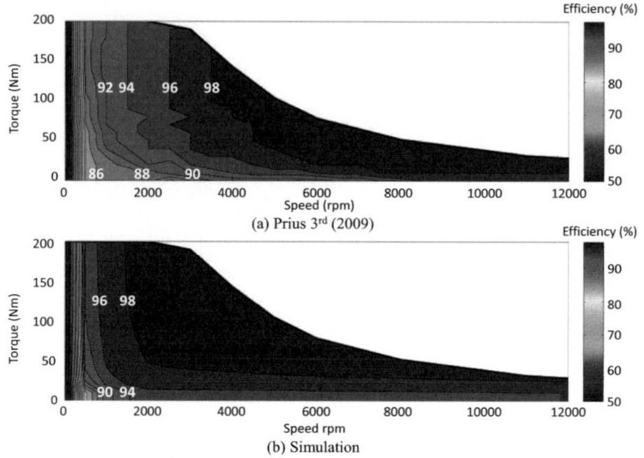

Figure 11: *Inverter efficiency*

3.3 Boost Converter Loss Modeling

Fig. 12 shows the equivalent circuit of the boost converter. A bi-directional DC-DC converter was used for the boost converter. Boost converter losses include the IGBT module loss, the loss due to the dc resistance of the smoothing capacitor, as well as the loss due to the dc and alternating current of the inductors. These losses are calculated as shown in (6) to (11) [9].

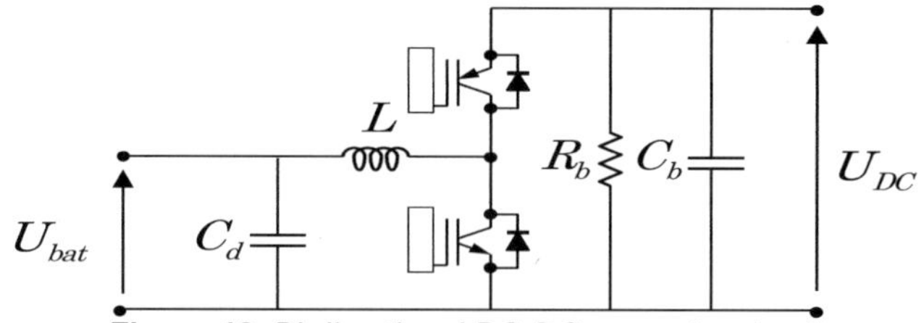

Figure 12: *Bi-directional DC-DC converter circuit*

$$P_{con}^{IGBT} = \alpha \cdot i_{out} u_{sat}^{IGBT} \left(\frac{1}{8} + \frac{m}{3\pi} \cos\varphi \right) \quad (6)$$

$$P_{sw}^{IGBT} = \frac{f_s}{2} (E_{on} + E_{off}) \quad (7)$$

$$P_{con}^{Diode} = (1-\alpha) i_{out} u_{sat}^{Diode} \left(\frac{1}{8} - \frac{m}{3\pi} \cos\varphi \right) \quad (8)$$

$$P_{re}^{Diode} = f_s E_{re} \quad (9)$$

$$L_C = (i_{out})^2 \, ESR + \left(\left(\frac{1}{\alpha} - 1 \right) i_{out} \right)^2 ESR \quad (10)$$

$$L_L = i_L^2 \cdot DCR = \left(\frac{i_{out}}{\alpha} \right)^2 DCR + U_{out}^{1.8} \cdot C \quad (11)$$

where,

α : current rate

DCR: DC resistance (Ω)

U_{out} : Output voltage (V)

C: Coefficient related to frequency, number of coil turns, etc.

The parameters of the IGBT modules and the like were set to the same as those used in the inverter loss modeling. However, the dc resistance of one smoothing capacitor was set to 5 micro-ohms (C_d) and 2 micro-ohms (C_b). The inductor dc resistance was set to 0.12 ohms, and C was assumed to be 0.001.

Fig. 13 shows the simulation results. The trends of the measured and estimated values are a close match [6].

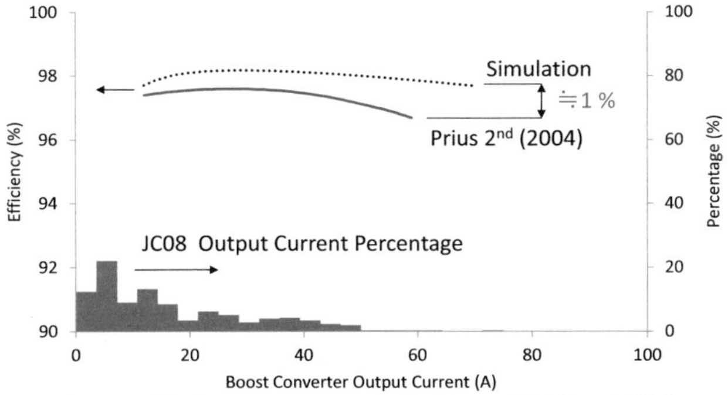

Figure 13: *Boost converter efficiency (233V→500V)*

4. Simulation Results

Table 1 shows the specifications of the PHV used in the study. The current estimation method and the inverter and boost converter loss models described above were incorporated into the simulator to calculate the loss of each electronic component for one JC08 test cycle. Fig. 13 shows the results.

***Table 1:** Vehicle specifications*

Component		**Value**
Curb Weight		1500 kg
Engine	*Max. Power*	73 kW
Motor	*Max. Power - Voltage*	60 kW - 650 V
Generator	*Max. Power - Voltage*	30 kW - 650 V
Battery system	*Voltage*	207 V
	Capacity	4.4 kWh

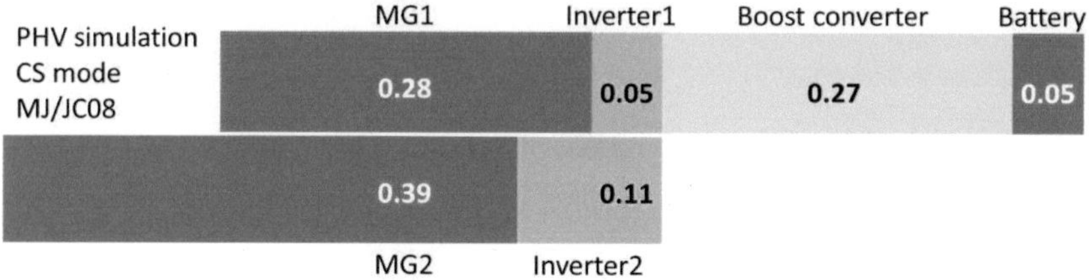

***Figure 14:** Motor system loss*

5. Conclusion

Detailed modeling of inverter and boost converter loss in a PHV motor system was carried out and the following results were obtained.

(1) Estimation of motor system current
The developed modeling method is capable of estimating the internal current of an IPMSM.

(2) Inverter loss modeling
This method was used to model the loss of an inverter comprising six IGBT modules and three smoothing capacitors. Although the overall estimated efficiency was on the high side, the trends closely matched each other.

(3) Boost converter loss modeling
This method was also used to model a DC-DC converter considering the loss of the IGBT modules, smoothing capacitors, and inductors. The loss simulation results showed a close match with actually measured data.

References

[1] Y. Sumida, A. Nomura, W. Yang, Y. Kamiya, Y. Daisho, and K. Morita, "Analysis of adverse effects on vehicle performance due to battery deterioration installed in BEV and HEV," 2nd Eur. Electr. Veh. Congr., pp. 1–8(USB), 2012.

[2] T. Kaneko, A. Nomura, W. Yang, Y. Daisho, Y. Kamiya, N. Sawada, et al., "No TiOptimization of engine control methods for range extender-type plug-in hybrid vehiclestle," 27th Int. Electr. Veh. Symp., pp. 1–8(USB), 2013.

[3] C. Mansour and D. Clodic, "Dynamic modeling of the electro-mechanical configuration of the Toyota Hybrid System series/parallel power train," Int. J. Automot. Technol., vol. 13, no. 1, pp. 143–166, Oct. 2011.

[4] T. Burress, S. Campbell, C. Coomer, C. Ayers, A. Wereszczak, and J. Cunningham, "Evaluation of the 2010 Toyota Prius hybrid synergy drive system," 2011.

[5] B. Bae, N. Patel, S. Schulz, and S. Sul, "New field weakening technique for high saliency interior permanent magnet motor," 38th IAS Annu. Meet. Conf. Rec. Ind. Appl. Conf. 2003., vol. 2, no. 2, pp. 898–905, 2003.

[6] R. H. Staunton, C. W. Ayers, L. D. Marlino, J. N. Chiasson, and T. A. Burress, "Evaluation of 2004 Toyota Prius hybrid electric drive system," 2006.

[7] J. Itoh and T. Oguru, "Evaluation of total loss for an inverter and motor by applying modulation strategies," Proc. 14th Int. Power Electron. Motion Control Conf. EPE-PEMC 2010, pp. 21–28, Sep. 2010.

[8] "6MBI300V-120-50," Fuji Electric Systems Co. Ltd. [Online]. Available: http://www.americas.fujielectric.com/sites/default/files/6MBI300V-120-50.pdf. [Accessed: 11-Nov-2013].

[9] M. Nikzad and A. Radan, "Accurate loss modelling of fuel cell boost converter and traction inverter for efficiency calculation in fuel cell-battery hybrid vehicles," Power Electron. Drive Syst. Technol. Conf., pp. 218–223, 2010.

10 Safe test bench equipment for automotive Lithium-ion batteries

Dr.-Ing. **Thomas Forchert**, Celono, Berlin, Germany, Dipl.-Ing. **Heiner Schmeck**, CTC cartech company GmbH, Boeblingen, Germany

Abstract

The physical implementation of lithium-ion batteries in test beds for combustion engines produces new challenges regarding the safety of personal and equipment. Long term start-stop or overload tests can set the lithium-ion battery in fire. The fire cannot easily extinguished by spraying with water. The article describes a new technical solution which avoids fire or avoids the spreading of fire to other components in the test bed. A new application for the building permit or an enlargement of the existing permit is not required.

Kurzfassung

Der Einsatz von Lithium-Ionen Batterien in Motorprüfständen stellt neue Anforderungen bezüglich der Sicherheit und insbesondere des Brandschutzes. Überlastungsprüfungen im Prüfstand oder Dauerlaufversuche im Fahrzeug können zu einem Brand der Lithium-Ionen-Batterien führen, welcher mit den üblichen Löschmitteln nicht sicher zu bekämpfen ist. Für die Erprobung der neuen Starter-Batterien sowie der Batterien für Hybrid- und Plugin-Fahrzeuge wächst der Bedarf an Prüfstandsläufen aktuell stark an. Der Artikel beinhaltet die Beschreibung eines völlig neuen Systemkonzepts zur Verhinderung eines Brands bzw. dessen Ausdehnung. Die Änderung der baulichen Betriebsgenehmigungen für bestehende Prüfstände ist dafür nicht erforderlich.

1. Introduction

Since lithium-Ion batteries come along with significant advantages concerning energy and power density they became preferred energy storage devices in the area of mobile electrical power supplies. Less weight and volume contribute to requirements concerning fuel consumption and additional space in terms of cross-section and feasibility of assembly.

Besides traction batteries, the carmakers increasingly introduce lithium-ion starter batteries in order to gain benefit from the advantages mentioned and replace lead-acid batteries, implemented so far. Hence, it is mandatory to obey some recommendations according the prevention of hazardous conditions that could occur while overcharging, in deep discharge situations and at high cell temperatures as well, induced by various circumstances.

2. Background and objectives

Nevertheless, a formation of dendrites could occur during operation and lifetime of a lithium-ion battery. Since dendrites will form the separating short circuits may increasingly damage the cells and finally lead to a thermal runaway, see also figure 1.

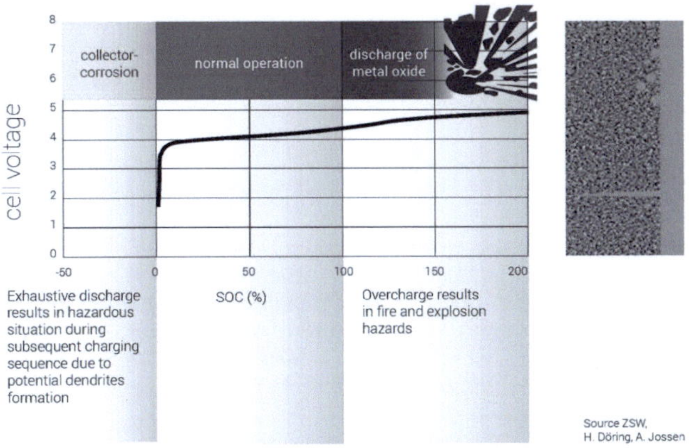

Figure 1: *Exhaustive discharging and overcharging results of a lithium-ion battery [ZSW, Doering/Jossen]*

As long as Li-ion batteries are implemented in cars and directly connected to the vehicle's infrastructure and control units concerned, they are considered as safe devices in a controlled environment. As soon as they are removed from this safe environment, they are immediately considered as hazardous goods and have to be treated accordingly. Obviously this will burden the operation of test stands for internal combustion engines, where starter batteries are operated similar to the "in-Vehicle" assembly so far and with no further safety measures required. The significantly increased hazard has to be mitigated and controlled, unless ICE tests should not be abandoned from the sensitive environment of OEMs and suppliers test factories.

3. State of research and technology

3.1 Lithium-ion safety issues

The high energy density of lithium-ion cells comes at a price. Manufacturing methods become more critical the denser the cells become. With a separator thickness of only 20-25µm, any small intrusion of metallic dust particles can have devastating consequences. Appropriate measures will be needed to achieve the mandated safety standard set forth by UL 1642. Whereas a nail penetration test could be tolerated on the older 18650 cell with a capacity of 1.35Ah, today's high-density 2.4Ah cell would become a bomb when performing the same test. UL 1642 does not require nail penetration. Lithium-ion batteries are nearing their theoretical energy density limit and battery manufacturers are beginning to focus on improving manufacturing methods and increasing safety. [1] batteryuniversity.com

3.2 Recall of lithium-ion batteries used in laptops

With the high usage of lithium-ion in cell phones, digital cameras and laptops, there are bound to be issues. A one-in-200,000 failure rate (5 ppm) triggered a recall of almost six million lithium-ion packs used in laptops manufactured by Dell and Apple. Sony Energy Devices (Sony), the maker of the lithium-ion cells in question, says that on rare occasions microscopic metal particles may come into contact with other parts of the battery cell, leading to a short circuit within the cell. Although battery manufacturers strive to minimize the presence of metallic particles, complex assembly techniques make the elimination of all metallic dust nearly impossible.

A mild short will only cause an elevated self-discharge. Little heat is generated because the discharging energy is very low. If, however, enough microscopic metal particles converge on one spot, a major electrical short can develop and a sizable current will flow between the positive and negative plates. This causes the temperature to rise, leading to a thermal runaway, also referred to 'venting with flame.' At 150°C the cell becomes thermally unstable, a condition that can lead to a thermal runaway.

During a thermal runaway, the high heat of the failing cell can propagate to the next cell, causing it to become thermally unstable as well. In some cases, a chain reaction occurs in which each cell disintegrates at its own timetable. A pack can get destroyed within a few short seconds or linger on for several hours as each cell is consumed one-by-one. To increase safety, packs are fitted with dividers to protect the failing cell from spreading to neighboring cells. [2] batteryuniversity.com

3.3 Carriage of Lithium batteries suspected in some accidents

An **Aug. 4, 2004**, incident in Memphis, Tenn., involved secondary batteries that were prototypes for an electric car. The batteries were packed in a container. The loaders smelled smoke and lowered it to the ground. When firefighters opened the container, the batteries burst into flame.

On **Feb. 7, 2006**, a UPS DC-8 caught fire while in flight and landed at Philadelphia International Airport. The three crewmembers evacuated safely, but the airplane was destroyed. While secondary lithium batteries were installed in several electronic devices shipped on that flight, the NTSB determined “No batteries were found that exhibited any damage identifying a source of ignition, nor could any determination be made that secondary batteries found in the debris had been subject to recalls.”

The fatal crash of a UPS Boeing 747 in Dubai **on Sept. 3, 2010**, was one of the subjects addressed by NTSB vice chairman Christopher Hart during testimony on April 7, 2011, to a Congressional committee about hazmat transportation. A preliminary report by the United Arab Emirates noted that although no hazardous materials were declared on that flight’s manifest, investigators reported there were on board “at least three shipments of lithium-ion battery packs that meet Class 9 hazardous material designation,” according to Hart.

The NTSB report 82 aviation incidents **between 1996 and 2007** involving all battery types. Of those, 14 involved secondary lithium batteries and 13 primary types. The 14 incidents included seven with items carried on cargo-only aircraft, six involving checked and carry-on baggage and one with cargo carried on a passenger aircraft. The passenger aircraft incidents included two in flight and five before takeoff (three on board, two at the airport). Of the seven cargo incidents, four occurred postflight, two before the cargo was loaded and one in flight, according to the NTSB. [3] chapt. 3.3. Matt Thurber, Aviation international news, 2012

Jan 7, 2013, a lithium ion battery aboard a Japan Airlines 787 parked at Boston's Logan International Airport lithium battery caught fire. Massport Fire Chief Bob Donahue said it appeared that a battery that powers the jet's electrical systems when the engines are turned off apparently had exploded. According to the National Transportation Safety Board, a battery in the auxiliary power unit suffered severe fire damage.

On **Jan 16, 2013**, a battery incident that forced an emergency landing in Japan by an All Nippon Airways 787. ANA 787 flight had an issue with its main battery only 15 minutes into a 90-minute flight. After 40 minutes, a burning smell made its way into the cabin and cockpit, and the plane made an emergency landing.
Those problems prompted the FAA and other authorities to ground all 787s for more than three months. The planes began flying again after Boeing changed the battery system, adding a tougher box to hold the battery and measures to contain any short-circuit or fire.

Figure 2: *JA829J at Boston Logan [nytimes.com]*

Jan 14, 2014, Battery problems resurfaced on Boeing's 787, after gas was discovered coming out of a battery on a plane parked in Tokyo. Boeing said the problem on a Japan Airlines 787 was discovered during scheduled maintenance. No passengers were on board. The company said it appears that a single battery cell "vented," or released gas. [4] Associated press

3.4 The Investigation of a Lithium-Ion Battery Fire Onboard a Boeing 787 by the US National Transportation Safety Board

[5] Joseph M. Kolly, Joseph Panagiotou, and Barbara A. Czech of the National Transportation Safety Board (NTSB)

The results of the examinations at the NTSB Materials Laboratory, with the results from radiographic examinations, enabled the NTSB to make public release of an initial set of findings earlier this year. The examinations revealed multiple cell failures within the battery, as evidenced, in part, by mechanical deformation and bursting of the vent discs. **This condition led the experts to conclude that the battery experienced a thermal runaway in which the failure of one battery cell cascaded to other neighboring cells within the battery assembly.** The initial failure was determined to be an internal short circuit in cell 6. This finding was supported, in part, by observations that cell 6 was located in the area of greatest thermal and mechanical damage. Additionally, clear evidence of internal short circuits was found within the electrode windings of cell 6.

Work continues to determine the cause of the internal short circuit. As of this writing, mechanical damage and external electrical short circuits of the battery have been ruled out as factors in the battery failure.

Figure 3: *Burned out Lithium-ion battery [NTSB]*

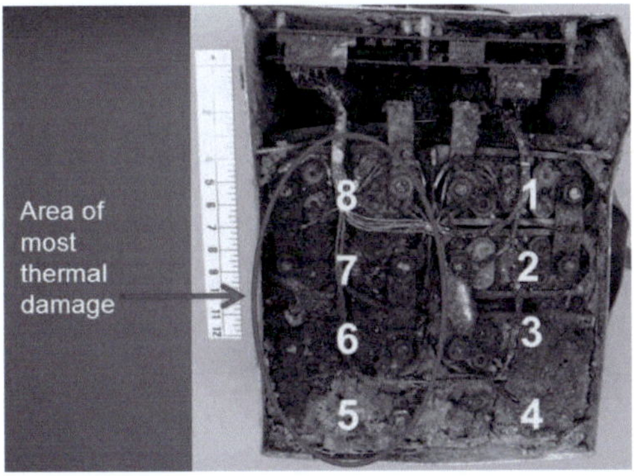

Figure 4: *The source of the fire [NTSB]*

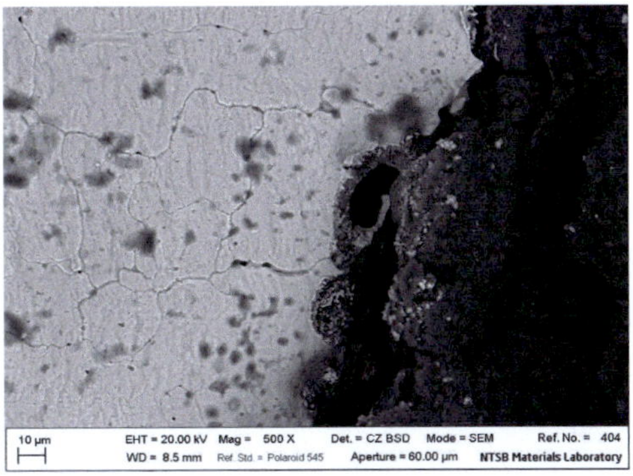

Figure 5: *Grain boundary decohesion near a foil hole, cell 6. [NTSB]*

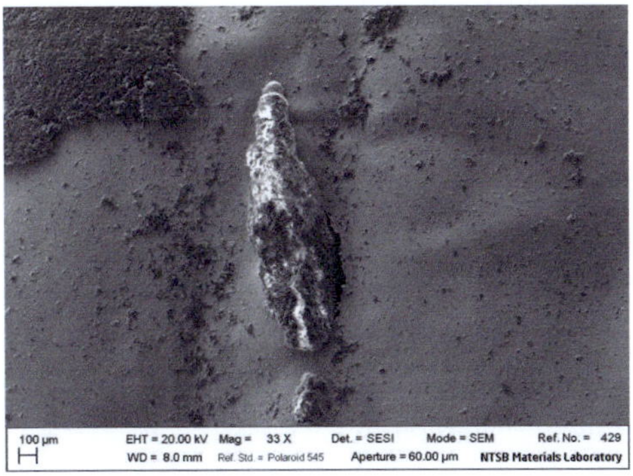

Figure 6: *Aluminum lump projecting through the bottom of a copper foil wrap, cell 6 [NTSB]*

3.5 Rule Changes for Battery Transport by air

In response to NTSB recommendations regarding lithium battery fires, the Pipeline and Hazardous Materials Safety Administration (PHMSA) noted that new rules that it has proposed would "require inaccessible rechargeable lithium-ion batteries to be transported in compartments or freight containers equipped with an FAA-approved fire-suppression system or in FAA-approved fire-resistant containers." The NTSB found that this response was acceptable, pending issuance of the final rule. But that applies to secondary (rechargeable) batteries. According to the NTSB, "current FAA-approved suppression systems are ineffective in preventing fires involving primary lithium batteries. Therefore, we urged the PHMSA to explicitly require that all shipments of these batteries be transported in FAA-approved fire-resistant containers until a fire-suppression system that is proven to be effective becomes available."

There is much more going on behind the scenes on the problem of lithium batteries carried as cargo, including FAA research, PHMSA rulemaking, NTSB concerns about growing numbers of lithium battery-related aviation incidents and advisory information disseminated to the aviation industry.

3.6 Available fire agents

Different fire agents are available to extinguish lithium- battery fires, see Figure 7 [Celono]. Celono decided for the glass granulate technology. The granulate bases on expanded glass, so-called foamglass. Celono prefers the product *EXTOVER®*.

		Glass gran-ulate	Ver mi-kulit	Min-eral wool	Wa-ter	Dry ice
Specific Density	Kg/m3	**250**	220	80	1000	1500
Specific heat capacity	kJ/kg K	**0,7**			4,2	
Thermal conductivity	W/mK	**0,07**			0,6	
Electrolyt-Sorption	l/kg	**2,0**			yes	no
Approuved as fire agent		**yes**	no	no	yes	yes
Reusability		**yes**	yes	no	no	no
Automated deloading		**yes**	yes	no	yes	yes
Recyclebility of battery		**yes**	yes	yes	no	yes
Aktive over years		**yes**	yes	yes	yes	no

4. The new approach

4.1 How works the expanded glass granulate?

First we demonstrate the result of a thermal runaway of a Li-ion pouch cell 3,7V/55Ah using no fire agent. The hot electrolyte vapor exploded in the air.

Figure 8: *Thermal Runaway, venting with flame no fire agent used [Celono]*

The following picture delivers the testimony of the expanded glass function in the situation of a thermal runaway with venting. The test has been delivered in a closed case. On top of the picture you see the exit point of hot electrolyt vapor (280 °C). The middle of the picture shows the electrolyte condensate area of about 60mm (darker colored).

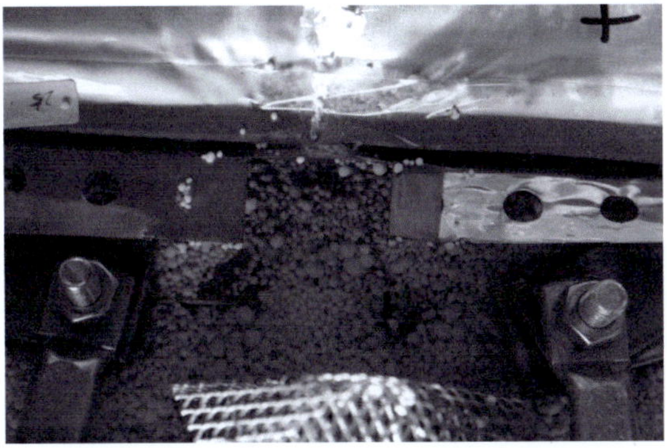

Figure 9: *Thermal runaway with venting, special expanded glass granulate as fire agent,* ***no flame*** *[Celono]*

4.2 The new method for the carriage of damaged Lithium-ion batteries

The new method is approved by the BAM - Federal Institute for materials research and testing at sept 2013. The following text has been published under D/BAM/GGVSEB (ADR)/ 2.2-275/13 [6]

For the carriage of damaged lithium-ion cells and batteries of UN 3480 by the road
"Within a rigid outer packaging made of metal of suitable size a layer of a special hollow glass granulate (expanded glass) with a thickness of at least 200 mm shall be filled. On that layer of this hollow glass granulate two lithium-ion batteries as a maximum should be placed. Between the side walls of the outer package and the lithium-ion batteries as well as in between the two lithium-ion batteries a free space of at least 150mm shall be left. This free space shall be filled with the special hollow glass granulate. Thus the lithium-ion batteries are surrounded from all sides by a layer of hollow glass granulate with a thickness of at least 150 mm. The lithium-ion batteries shall be covered above with a layer of special glass granulate with a thickness of at least 200mm. This special glass granulate is used as inert, non-conductive and non-combustible absorbent and cushioning material. It shall be assured that 1.5 times of the amount of liquids which could leaked will be absorbed by the special hollow glass granulate. The melting point of the special glass granulate shall be more than 1000 °C."

5. Test bench equipment

The mitigation concept prepared and applied right now is subdivided into three different mitigation layers. First and foremost it was important to create a protective environment for the operation of Li-Ion batteries where the units work insulated from the test stands surroundings. A welded steel container houses the battery as a first approach. The compartment fully encloses the Li-Ion unit. In addition an expanded glass granulate fills the container and completely covers the battery.

Dimensions of the container and volume of the granulate are determined from the hazard load of the batteries to be operated. Furthermore the granulate absorbs a leaking electrolyte as the case may be. If, in the worst case, a thermal runaway occurs, the granulate cools the hot vapor and prevents a flaming. The technical solution is approved according BAM (see chapter 4.2) moreover.

We call this mitigation level the **Fire Suppression Layer** as the highest and last protection measure at the same time.

Basically, mitigation starts at the battery system side where LIN-bus access allows the acquisition of multifarious data. Voltages, current, temperatures and logical data on the charging and discharging process can be obtained on the cell level if required. Thus, a first diagnosis of the battery's health provides an insight into the system's load and a deduction of a potential upcoming hazard at the same time.

In case of hazardous incidents being detected the battery is to be set free of charge and the according tests should be terminated prior a potential further escalation.

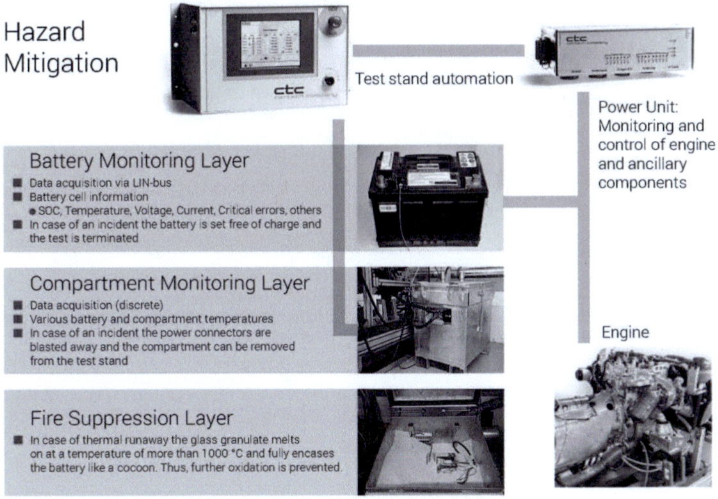

Figure 10: *Hazard mitigation concept [CTC cartech company]*

The **Battery Monitoring Layer** is the first and basic level of hazard mitigation and should not be skipped, rather than it could provide sufficient protection for the entire test set-up and prevent from further escalation. Availability of data connections and thus, access to battery information could be limited, denied by the manufacturers, disrupted or simply not provided through the battery management system.

Hence, on the next level we introduced the **Compartment Monitoring Layer**, taking care of already damaged batteries and ongoing abnormal processes. Therefore various temperatures are acquired from battery and compartment as well in order to permanently deduct temperature profiles within the container. In case of incidents detected, the power connectors are blasted away and the entire compartment can be removed from the test stand quickly.

Figure 10 shows the interaction of the 3 mitigation layers: A CAN-controlled Power Unit monitors and controls the ICE including the ancillary components, one of which is the battery system. Data from the battery are acquired and conveyed the overlaying automation unit. The automation box on the other hand monitors and acquires data from the container. The automation unit is the overlaying master system for the entire test and monitoring process, provides various data acquisition capabilities and can be controlled remotely. Thus, a safeguarded operation and monitoring of ICE tests including the operation of Li-Ion batteries could be implemented. Data aggregation, analysis and interpretation services complement the recent test stand concept and provide full flexibility for further extensions.

Figure 11: *Operational test set-up. [CTC cartech company]*

6. Summary and outlook

New lithium-ion and post lithium technologies with much higher energy density will occur on the market. The technological progress combined with cost reduction efforts has the tendency to produce increasing risks. The risks can be controlled in the test environment, the transportation and at least in the onboard systems.

As long as we work in the pioneering situation of making batteries mature enough for (auto)mobile applications, we have to cope with all hazards and uncertainties occurring during the development process. In a bottom-up approach we have to develop methods and tools to mitigate risks and to foster the integration process.

CTC focuses a.o. on the development of high tech test beds, methods and tools to approve ancillary ICE components, as well as lithium-ion batteries before the market launch. Moreover CTC operates test stands for electrical drive trains which are enhanced permanently.

Celono works on technical solutions to suppress fire in the onboard environment. The lithium battery technology will be liberated from fearing smoke and fire in the future.

Literature

[1-2] www.batteryuniversity.com, 2014

[3] Matt Thurber, Aviation international news, 2012

[4] Associated press, 2014

[5] Joseph M. Kolly, Joseph Panagiotou, and Barbara A. Czech of the National Transportation Safety Board, 2013

[6] BAM Federal Institute for materials research and testing, Document D/BAM/GGVSEB (ADR)/ 2.2-275/13, 2013

11 Intelligent Charging Infrastructure Research Platform

Dipl.-Ing. **René Unger**, Dipl.-Ing. **Martin Heinrich**, EA Systems Dresden GmbH, Dipl.-Ing. **Ronny Petersohn**, Technische Universität Dresden, Dresden, Dipl.-Ing. **Andreas Unger**, IAM GmbH, Dresden, Germany

Abstract

A new research platform for intelligent vehicle charging and management (ICIRP) is currently implemented in cooperation of IAD TU Dresden and IAM GmbH and EA Systems Dresden GmbH.
The paper shows the concept and essential features of this platform which combines real-time data, diagnostics and interaction between vehicles, charging infrastructure and user via Mobile Internet devices. Additionally the possibilities of coupling the system to a Modelica infrastructure simulation are shown.

Kurzfassung

Von den Partnern IAD der TU Dresden, IAM GmbH und EA Systems Dresden GmbH wird gegenwärtig eine neue Forschungsplattform für Fahrzeugladung und Verwaltung entwickelt. Das Paper beschreibt die Anforderungen an ein solches System, dessen Realisierung und die Möglichkeiten der modularen Erweiterung. Im Fokus stehen dabei Echtzeitdatenerfassung, Diagnose sowie die Interaktion zwischen Fahrzeugen, Ladeinfrastruktur, Webplattform und Nutzern. Weiterhin werden die Möglichkeiten der Kopplung an eine Infrastruktursimulation beschrieben.

1. Introduction

Getting a substantial number of electrified vehicles to the road is a major issue for future mobility concepts. While all major car manufacturers have hybrid or battery electric vehicles in their product lineup, the required support infrastructure in Germany still consists of prototypes and flagship projects.
The main challenge for electrical vehicle usage is range. This can be compensated with a good usage strategy (know when to go where) and a reliable charging infrastructure (easy to find, cable compatibility, not occopied). Today this is more a future vision with the cable drum in the trunk (CDT-Solution) being the reality.
There are two main strategies for charging applications. One strategy is to charge with the maximum possible power. The charging time is seen as an interruption to the mobility demand of the user. This strategy is similar to the classic refueling scenario of the conventional gas- or diesel driven car. This fast-charging is complicated by the immense hardware demand with thick cables, transformers, waste heat dissipation and the power demand to the public grid.
The other strategy is to charge for a longer time with less power. To make this suitable for the user, low power charging must correspond to planned duration of parking

and the planned route. This has to be synchronized with the daily schedule. For long distances a better suited vehicle (i.e. diesel driven) needs to be ordered. It is easy to see, that the demand on charging infrastructure is much lower, but a lot of extra planning and informational infrastructure is needed.
To approach these challenges, the Institute of Automotive Technologies of Dresden Technical University, EA Systems Dresden GmbH and IAM GmbH work together in developing an intelligent charging infrastructure. The new charging station is a capable platform to research usage scenarios, fleet management, grid interaction and renewable power charging strategies in current research projects like EnMover, ECityRouting and EMiD.

2. Requirements

Previous projects like "Residence and Mobility" (c.f. Figure 1) showed the demands for such an intelligent charging solution in research & development projects. The main objective is to provide an easy to use and well instrumented interface to vehicle charging. That way the engineer or scientist can concentrate on the research topic, for example the fleet management algorithm.

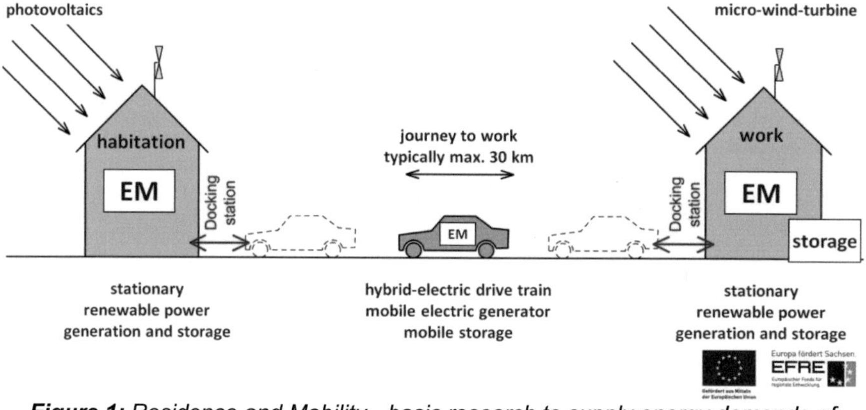

Figure 1: *Residence and Mobility - basic research to supply energy demands of home and commuting with local renewable energy [1]*

An important requirement is the capability for dynamic power switching based on external inputs. Typical inputs are available renewable power or the limit of the lateral connection line. This is especially important to research fleet charging strategies.
Additionally an instrumented charging station needs to support a complete set of measuring devices regarding power metering, environment sensors or grid frequency analysis. This allows evaluation of grid perturbations.
The system needs to be extensible to integrate modules which are not standard. Therefore connectivity to other management systems (machine2machine) is as important as standard user interaction via web platform and mobile internet devices.
For advanced testing the platform needs to support real-time monitoring and remote control with quick reaction times for user interaction.

If something fails, most functionality should also be available even if the station is disconnected from its uplink. This is especially important in security enhanced research networks where connectivity is often a bureaucratic issue.
In fleet testing, the price per charging point is a major issue for feasibility. Since the instrumentation and internet connection to user require a lot of cost intensive hardware this conflict is alleviated with a master-slave concept, where a more expensive central station support a number of 'dumb' slave outlets.

3. easyCharge2 - flexible and modular Platform

The new charging infrastructure implements the mentioned requirements to provide an out of the box solution, so researchers can concentrate on their objectives.
The system is made up out of four levels: internet platform, local management PC, PLC & communication devices and smart meters & scientific instrumentation.

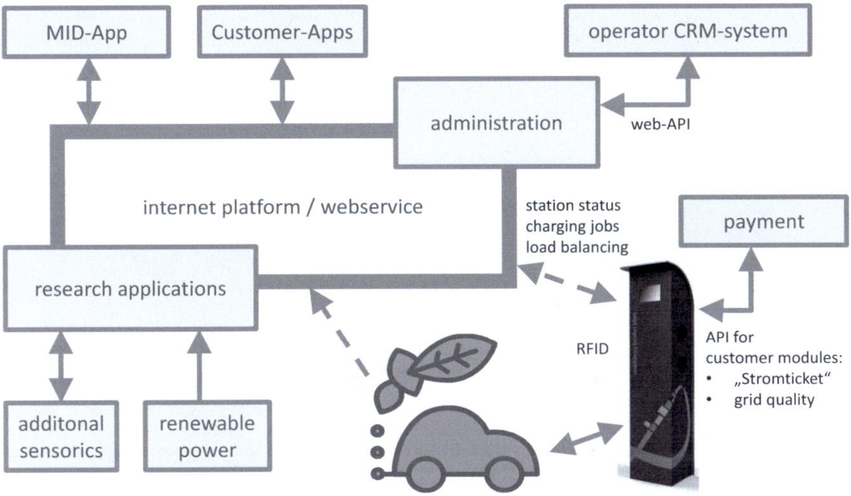

***Figure 2:** easyCharge2 plattform concept*

The internet platform provides a monitoring and diagnostic interface. With this the users can view their present and historic charging jobs with detailed data like current and maximum power, charged energy and renewable share as well as status, i.e. 'finished'. Additionally they can change the current charging strategy.
The administration interface is designed to manage user clearance for stations with RFID, PIN or similar authentication. The users and stations are managed in flexible groups, for example it is possible for User A to stop and unlock the charging of User B if they share the respective group. This is especially important for small vehicle fleets where employees of a certain department share a set of vehicles but should not 'play' with the vehicles of a different department.
The platform also runs regular housekeeping services like diagnostics and error reporting, email messaging the user on finished jobs or building statistics and forecast

data, i.e. grid quality. Monitoring is set up to work for regular logging interval of one minute, faster is possible if performance and disk space are no issues.
For connectivity to external systems different Rest-Services are provided for user management and accounting, monitoring data acquisition as well as remote control and mobile device apps. Custom extensions can be plugged in either via Rest-Service or directly into the housekeeping sequence.
Within the charging station, the local controller software runs on an industrial PC platform with Linux Operating System. The software and graphical user interface can run fully offline. Synchronization of user management and historic data starts if connectivity to the server is established. Since remote access and updates of the software are of great importance in development projects, this can be easily done, without any router configuration, by enabling the integrated tunneling option.
The local charging controller provides the customizable user interface (content and design) as well as authentication and management of the charging jobs. Depending on the strategy, the available renewable power, strength of grid connection and priority, the current for the different charging outlets is set.
Custom extensions can be installed either as standalone application in the Linux environment or as Java module directly within the charging controller. Additional options are the local webserver or enabling direct charging point control via Rest-Service. Since all controller and database (PostgreSQL) software is platform independent, the usage of Microsoft Windows as operating system is also possible.

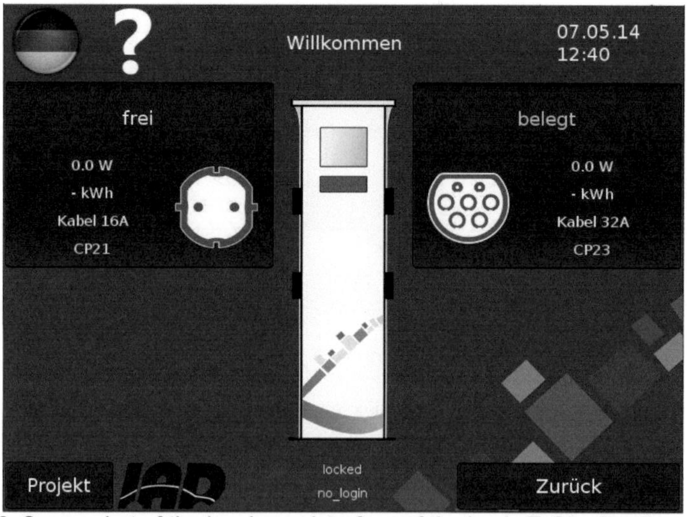

***Figure 3**: Screenshot of the local user interface of the basic E-City-Routing charging station*

The component level consists of the IEC vehicle communication devices (Phoenix Contact EV Charge Control), RFID module, motion sensor and smart meters measuring power, energy and phase currents. Optional PLC modules for additional digital and analog IOs as well as a grid quality monitor are available. The components communicate to the management pc via internal LAN or serial bus (MBUS).

The advantage of this modular structure is that the charging point configurations is flexible and can be configured to the project needs. Additionally it is possible to integrate custom modules at each level of the platform.
Currently outlets as earthed safety socket (DIN49440) or as EN 62196-2 (Mennekes) are supported. A special option is fuse switching, so 16 A cables as well as 32 A cables are safely supported at the same EN 62196-2 outlet.

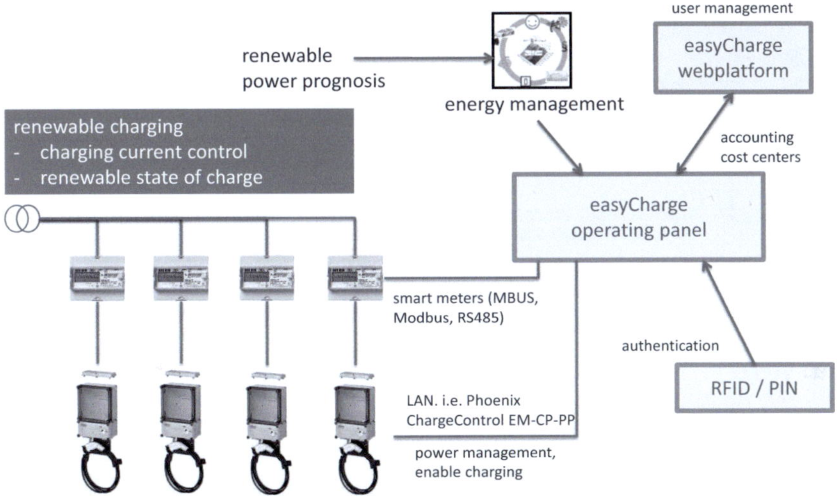

***Figure 4**: System layout for a master-slave charging concept with a central control unit and several wallboxes*

4. Modelling Infrastructure Interaction

Modelling and simulation of the charging process as well as the surrounding power infrastructure is an important part of the research process. ICIRP connects to simulation models in different ways.
In the first step, charging algorithms were simulated using the SimulationX platform with Modelica models for vehicle and battery developed at the IAD Dresden as well as infrastructure and smart grid models taken from the Green Building Library developed by EA Systems.
In the simulation, the charging station control algorithm can be placed into a local grid, building energy system or an urban district. The setup is used fort testing the interaction to other smart grid components like photovoltaic generators, combined heat and power units, heat pumps, battery storage or user behavior. This is especially interesting if the peak power from the grid is limited or if other controllers, like a central energy manager, interact with the charging station. A typical example is a situation where surplus power could be used either for faster charging or for the heat pump (Power2Heat storage).
In the second step the implemented controller hardware and software can be tested in the Loop (XiL) at the same infrastructure model. For this a connector library based on the FMI co-simulation standard was used. This tool generates a Java library for

the software under test well as an FMU which is imported into the simulation environment. The two are connected by a transparent network interface, so controller and simulation can run on different platforms.
In the third step the acquired monitoring data from the charging processes can be used to improve the models.

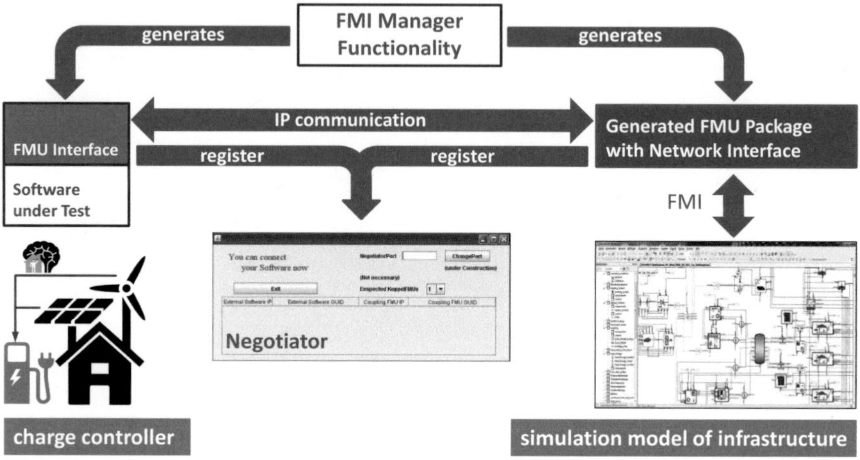

Figure 5: *FMI-connector overview: software under test and simulation model [2]*

5. Conclusion and Future Work

The described charging station research platform makes it possible to test new technologies within a functional platform. This is used for current emobiliy research projects at Dresden Technical University. The platform is available for other projects as well. The modular system can be extended for the individual project needs.
Current development addresses new communication systems like GreenPHY compatibility, integration of OTX Services and improved prediction of drive cycle power demand.
Options we like to address in the future are integration of DC charging points as well as new charging algorithms integrating model predictive control for higher renewable energy share and reduction of grid asymmetry.

Literature

[1] Unger, R., et al., 2014, EA EnergieArchitektur GmbH, final project report „Wohnen und Mobiltät“

[2] EA Systems Dresden, ITI GmbH, information material FMI Manager

12 Multi-physics modeling approach for energy efficient drive solutions

Dipl.-Ing. **Salim Chaker**, Dipl.-Ing. **Christian Kehrer**, ITI GmbH, Dresden, Germany

Abstract

Besides the common requirements for more and better results in less time, the automotive industry in particular needs virtual design and testing methods in order to reuse drivetrain models and the corresponding data throughout the entire development process including offline and XiL scenarios. Such methods may also help to close the gap between merely theoretical simulation results and the actual driving experience by bringing the whole simulation to life. The presented approach tries to do just that. It embodies a modeling paradigm of non-causal forward calculations which forms the basis for a variety of applications. Model types from the various libraries with different levels of detail and physical functionality can be adapted to the analysis' objective. They allow for easy modifications or can be created from scratch, while standardized connectors ensure interchangeability for seamless cross-domain modeling. On the simulation level, results can be fed back into the global setup in real-time. This enables multiple tools to be used in one environment (e.g. SimulationX, Matlab/Simulink) and simulate the complex interactions of all factors involved (e.g. driver, driving cycle, operating strategies). The data exchange between these tools can be performed by using supported proprietary interfaces or universal standards, such as FMI, which are all based on generic C code. The sheer endless possibilities of such a complementary approach allow detailed analyses and energetic calculations on different modeling levels and across various domains (e.g. switching thermal management and elasticities on and off). It also facilitates the integration of models into additional infrastructure to evaluate mobility and energy management concepts as part of smart grid systems (BEMS, HEMS), for instance.

Kurzfassung

Neben allgemeinen Forderungen nach umfangreicheren und genaueren Ergebnissen in kürzerer Zeit bedarf gerade die Automobilindustrie virtueller Entwürfe und Testmethoden, um Modelle von Antriebssträngen und die damit verbundenen Daten im gesamten Entwicklungsprozess auch für Offline- und XiL-Szenarien wiederverwenden zu können. Derartige Methoden können zudem dazu beitragen, die Lücke zwischen rein theoretischen Simulationsergebnissen und dem tatsächlichen Fahrerlebnis mithilfe realitätsgetreuer Simulationen zu schließen. Der vorgestellte Ansatz versucht, dies anhand von akausalen Vorwärtsrechnungen darzustellen, und bildet damit die Grundlage für eine Reihe verschiedener Anwendungsmöglichkeiten. Modelltypen aus den verschiedenen Bibliotheken mit unterschiedlicher Detailtiefe und physikalischer Funktionalität können an die jeweilige Zielsetzung angepasst werden. Sie erlauben einfache Modifikationen oder können von Grund auf neuerstellet werden, wobei standardisierte Anschlüsse für eine domänenübergreifende Modellierung sorgen. Auf Simulationsebene können Ergebnisse in den Gesamtaufbau in Echtzeit übertragen werden. Damit können mehrere Werkzeuge in einer gemeinsamen Umgebung genutzt werden (z. B. SimulationX, Matlab/Simulink), um komplexe Wechselwirkungen aller beteiligten Faktoren zu simulieren (z. B. Fahrer, Fahrzyklus, Betriebsstrategien). Der Datenaustausch zwischen den verschiedenen Tools erfolgt über eigens dafür entwickelte Schnittstellen oder universelle Standards wie FMI, die allesamt auf C-Code basieren. Die schier unbegrenzten Möglichkeiten einer solchen komplementären Methode erlauben detaillierte Untersuchungen und energetische Berechnungen auf unterschiedlichen Modellierungsebenen über verschiedene Domänen hinweg (z. B. das An- und Ausschalten von Wärmemanagement oder von Steifigkeiten). Sie vereinfacht zudem die Einbindung von Modellen in zusätzliche Strukturen, um zum Beispiel Mobilitäts- und Energiemanagementkonzept als Teil von Smart-Grid-Systemen (BEMS, HEMS) zu bewerten.

1. Motivation

In today's world, limited resources in all areas of everyday life are like the sword of Damocles. That includes not only fossil energy sources or energy storage capacities from a technical point of view, but also the challenge for the human ability to deal with ever more complex correlations in a globalized world. In the same way, engineers face a variety of challenges in their daily routine during the development of energy-efficient products, such as vehicles or technical components. Because of the great diversity of such multi-domain systems, the impact of structural or parametrical changes can no longer be predicted intuitively. Consequently, comparative investigation, selection and testing of an appropriate system concept require more virtualization in the development process to keep a tight schedule. Ever-growing availability and performance of computing capacities make moving these design stages to earlier development phases possible. The real bottleneck, however, will be the user in the middle facing increasing difficulties to manage all that data.

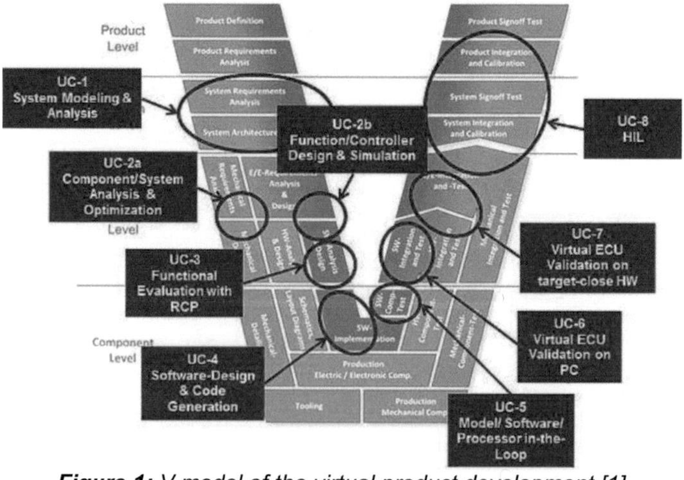

Figure 1: *V model of the virtual product development [1]*

It is therefore inevitable to elaborate a concept in order to ensure that simulation models and the corresponding data can be re-used throughout the entire development process including offline and XiL scenarios (comp. Figure 1).
To support engineers in an efficient and continuous virtual development process, the objective of this work is to demonstrate a general modeling approach based on a generic library concept with an open modeling language and standardized tool interfaces.

2. Modeling paradigm

For the modeling of technical systems, a variety of modeling methods and approaches are available. A basic classification can be made according to the effect relations between the model components. Here, a distinction is made between causal (signal-flow-oriented) e.g. Matlab/Simulink and non-causal (object oriented) e.g. Modelica® modeling. The properties of these two main modeling approaches are summarized in the **Table 1** for the two example applications, SimulationX and Matlab/Simulink.

Non-causal, object oriented modeling (Modelica®/SimulationX)	**Causal, signal-flow-oriented modeling (Matlab Simulink)**
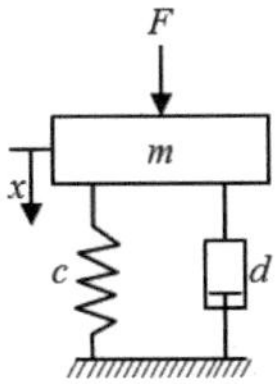 $\ddot{x} + \frac{d}{m}\dot{x} + \frac{c}{m}x = \frac{F}{m}$	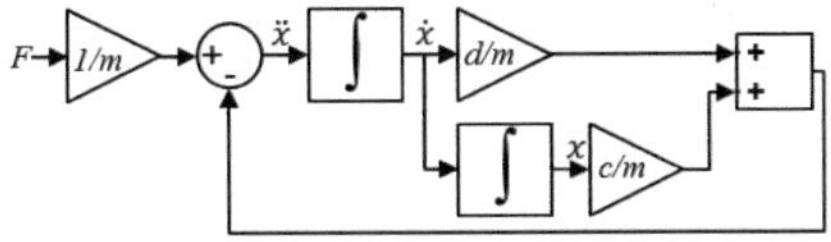
• Direct implementation of the differential equations describing the system • Free solving direction (non-causal modeling using Differential-Algebraic Equations) • Symbolic analysis and pre-solving of equation systems (order reduction), then numerical integration (→ reducing the computational effort) • Easy creation of complex systems by connecting via interfaces that are similar to the real system • Identical structure of model and real system (good model readability) • Class structure and inheritance (high reusability and extensibility)	• The implementation of the differential equations describing the system requires an additional reformulation to determine the solving direction (possibly remodeling of an inverse model required) • Numerical solution of the model block by block, no symbolic analysis and pre-solving of the equation system • Extensive creation of complex systems by means of manual resolving of the differential equation systems • High abstraction of the real physical system (poor model readability) • No inheritance or class structure (poorer reusability)

Table 1: *Property comparison of non-causal object oriented and causal signal flow oriented modeling*

Depending on the perspective of control engineering, the causal modeling formulates the functional relationships in the form of blocks with inputs and outputs. In this case, the direction of the action and thus the solving direction are always predefined. Since real physical systems rarely have a direction of action, it is often required to resolve the system of differential equations describing the systems. The limitation of causal modeling are inevitable, especially when modeling complex systems with differential-algebraic linked multiple components. Non-causal approaches are not subject to such restrictions, so even very large systems can be modeled intuitively. Here, instead of inputs and outputs, connectors are used as interfaces for flow variables (e.g. current, torque, force...) and potential variables (e.g. voltage, displacement, speed...). The overall equation system is based on the internal equation system of the components and the equations generated by the linked connectors (equality of potential variables, balance of flow variables). This declaratively formulated system of equations is transformed into symbolic form and pre-solved, so that a very time-efficient simulation can be performed.

3. Modeling concept

Within the scope of this article, only driving performance and fuel consumption will be considered. Analyzing vibration behavior and mechanical strength is not possible before choosing the basic powertrain concept. Nor will emissions be taken into account to keep the model complexity manageable. To estimate the mentioned driving aspects of hybrid powertrains depending on environmental conditions and driver actions, a library named "Hybrid Powertrain Package" was created by means of modeling and simulation environment SimulationX (cp. Figure 2).

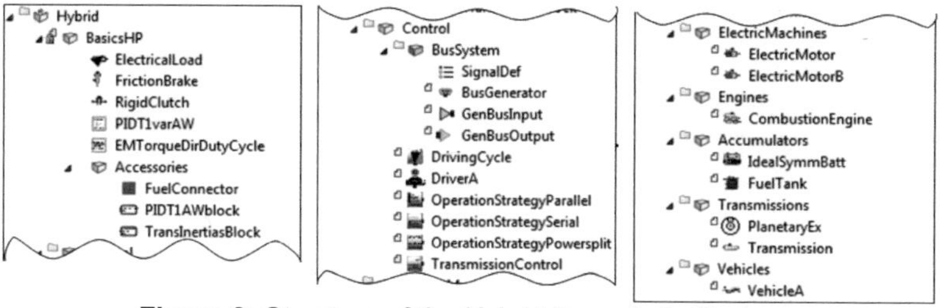

Figure 2: *Structure of the Hybrid Powertrain Package*

The library elements model the mechanical, the electro-mechanical and the electrical vehicle domains. They have universal connections that allow these elements to be used with other library elements such as thermal and hydraulic ones. The first library subpackage "BasiscsHP" includes basic elements such as brake, clutch, electrical loads and different controllers for any model. The sub-package "Control" contains all the components (bus system, driving cycle, driver, operating strategies and transmission controller) for stimulating and controlling the powertrain. Two electrical machines with two different degrees of model depth are found in the sub-package "ElectricMachines". These can be parameterized either as a DC or a synchronous motor. The first machine is based on the electrical equivalent circuit of an electrical machine. The second one is map-based. In addition, the library has a model of an internal combustion engine with a fuel consumption map and without consideration of the rotational non-uniformity (optimization of the computation efforts). The Library includes also a model of an ideal battery with charge /discharge characteristics, a transmission model with two to ten gears, representing AMTs or ATs and a longitudinal vehicle dynamics model with driving resistances and elastic drive shaft. All library elements have a bus interface for connections with the bus system. Thus, each component in the model has access to all result variables needed for the calculation of the output quantities. The calculated output quantities are also written to the bus, so they are available for all other model components. This connection approach has the following advantages:

- Central definition of all variables to be exchanged between the components
- Easier linking of model components
- Easier creation and extension of model types
- Improved model clarity, modeling flexibility and comfort.

The well-structured interaction of all components requires the definition of a stringent modeling method. For the modeling of both, longitudinal dynamics and powertrain, there are two methods – the quasi-static one and the dynamic one. Dynamical modeling, which was selected for the Hybrid Package and is shown in **Figure 3**, is based on a forward calculation.

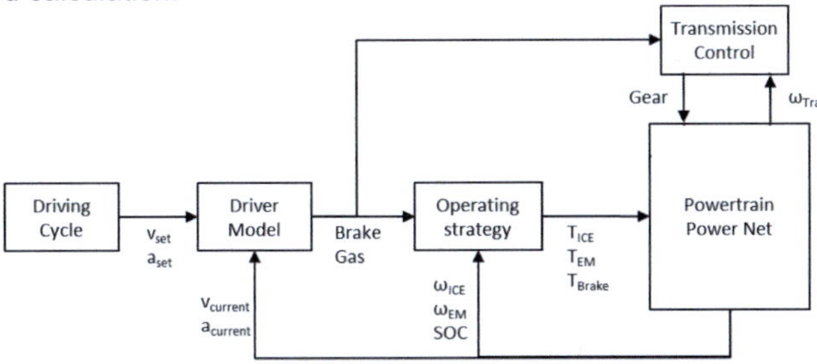

Figure 3: *Controller concept of the Hybrid Package*

A driver model calculates the overall drive torque needed in such a way that the difference between target and current velocity becomes as small as possible. The driver model receives the target velocity from the driving cycle, the current velocity from the vehicle/powertrain and further quantities if necessary. The calculated target torque is then passed on the operating strategy. This concept differs from the signal flow in a real vehicle, where the driver does not specify the target torque but operates acceleration and the brake pedal. Nevertheless, the described modeling concept is used, because of the reduced model complexity which helps to understand the system [1]. Consider the driver's behavior and driving states (SOC, maximum torque driving/braking states etc.), the operating strategy splits the target torque to the drive units of the powertrain (combustion engine, electric machines, brake) and ideally controls it in a way that the energy consumption is minimized. The structure of the powertrain that is to be controlled has a significant impact on the necessary output quantities and the functionality of the operating strategy. The developed package enables the user to model all kinds of powertrain structures. However, a general operating strategy serving all these structures is not possible. Instead, simple operating strategies for parallel, serial and power split hybrids are modeled and can be used as

templates for more complex variants. The optimization of such strategies is not part of this work though [2].
The gearbox is the only powertrain component which is not controlled by the operating strategy, because it is independent of the control of other components. Therefore, a dedicated gearbox controller calculates the appropriate gear from the angular velocity of the gearbox output shaft and the target torque. Although, in reality manual transmissions are controlled by the driver, this gearbox controller is necessary for both automatic and manual transmission models. In other words, the gearbox controller adopts a part of the driver's functionality. This leads to a consistent functional partition of the controller components, independent of the structural design of the powertrain. The described control concept can be divided into a control level and a physical level. The control level consists mainly of signal-flow-oriented elements (causal modeled) to control and stimulate the physical powertrain components. They are modeled as non-causal components (to enable bi-directional energy flows as in reality), interconnected by means of a physical interface and exchange their operating states with the control level via the bus system (see **Figure 4**).

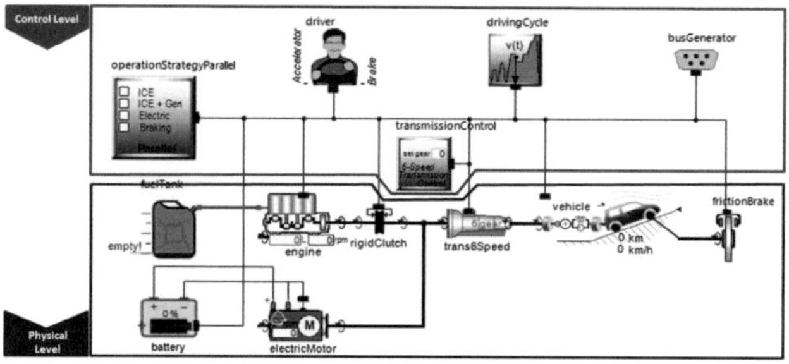

Figure 4: *Model structure of a parallel hybrid powertrain (Hybrid Package)*

4. Simulation approach

For modeling the plant powertrain, non-causal object oriented modeling is the preferred approach, whereas signal flow oriented modeling is the preferred methodology for the implementation of controller components of the powertrain (engine control, transmission control, operating strategy). Due to the good interoperability of the tools Matlab and SimulationX, a dual approach was chosen using the advantages of both tools. For the realization of such a simulation concept, model exchange or Co-simulation between the used tools is essential, including the following advantages:

- Providing a component model to OEMs: Here, a distinction between a black-box model and a white-box model can be made.
- Using models from other modeling environments that have certain characteristics and advantages, to increase the modeling quality and simulation efficiency.
- Reusability of models independent of modeling environment (FMI-standard) and Simplified way to test and validate model
- Implementation of distributed simulations for the coupling of models from different domains and to increase the available computing power per model.

The following figure summarizes the possibility for model exchange and Co-simulation between SimulationX and Matlab.

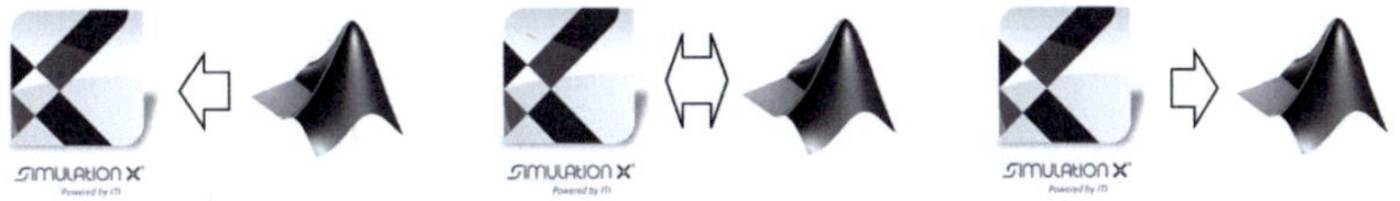

(a): FMI for Co-simulation (b): Co-simulation (c): S-function code-export

Figure 5: *SimulationX- Interfaces to Matlab/Simulink*

SimulationX can import source code generated by the Simulink coder as an FMU for Co-Simulation (see **Figure 5**(a)). The Simulink coder generates source code for a DLL which can be used in SimulationX. The DLL uses the fixed step solver of Simulink which was selected before the creation of the DLL. The simulation calculation runs separately in SimulationX and the generated DLL. The data between the DLL and SimulationX will be transferred at the sample time steps. This method is suitable for the integration of any operating strategy developed in Simulink into SimulationX. The Co-simulation-interface of SimulationX to Matlab/Simulink uses the TCP/IP protocol (see **Figure 5**(b) and **Figure 6**). ITI's SimulationX can take control over the simulation process as master or can be controlled as slave by the co-simulator. Both simulators can use their own calculation step-size between the communication steps. Values between the communication points can be calculated by linear extrapolation. Therefore, the previous value of each variable and the corresponding calculation step-size are included in the exchanged packages. This method is suitable for the realization of a distributed simulation on multiple simulation environments. The advantage of this method over the mere use of Simulink as a host environment is the individual choice of solution and integration algorithms for each sub-model, which has advantages with respect to the simulation time promises, especially with large, multi-physical systems (mix of systems with very different time constants). However, it is important to take into account the communication overhead for synchronization between the sub-models.

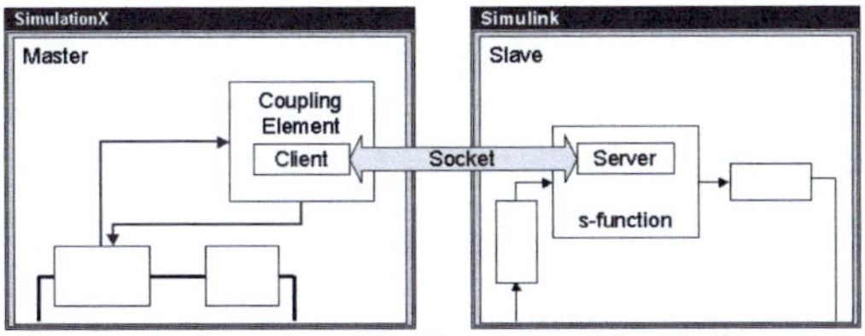

Figure 6: *Coupling ITI SimulationX with Simulink*

For the realization of offline simulations in Matlab/Simulink, the export of SimulationX models as S-function (system function) as a mex-file is suitable (see **Figure 5**(c)). This method is not suitable for our simulation scenario since the structure of the physical plant is no longer visible and customizable.

5. Generic toolchain application

To benefit from the advantages of each modeling paradigm (causal, non-causal) the control level of the parallel hybrid vehicle model (see **Figure 4** and **Figure 3**) was isolated and modeled in Matlab/Simulink (cp. **Figure 7**).

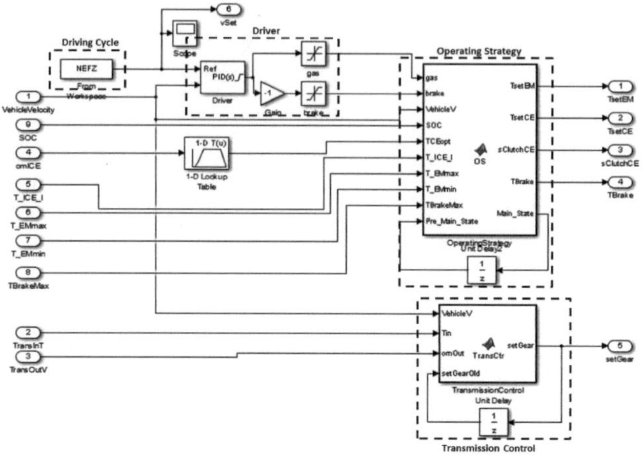

***Figure 7:** Control level modeled in Matlab/Simulink*

The control level modeled in SimulationX was used here as the starting point for the modeling in Matlab/Simulink. The operating strategy was implemented with the help of the state-chart designer of SimulationX. For its implementation in Matlab/Simulink both state flow and Simulink functions (m-files) can be used. The main focus of it is to keep the battery's state of charge (SoC) within defined upper and lower limits. This requires a parameter for the hysteresis to avoid a permanent switchover between different operating points. Such a hysteresis is also used for the target driving torque. For an effective control of the engines, the operating strategy needs information about their speed-dependent performance capabilities (see **Figure 8**).

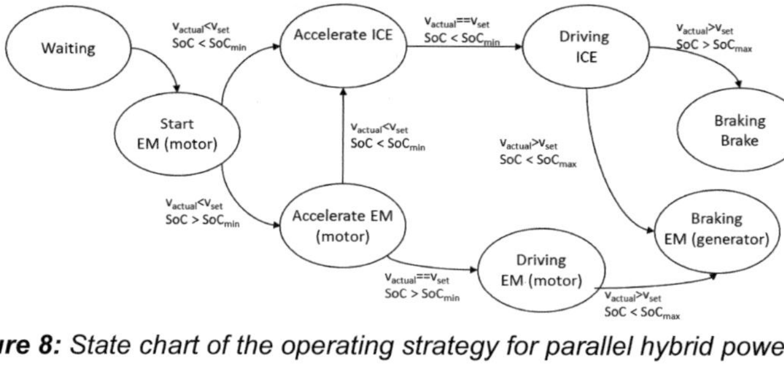

***Figure 8:** State chart of the operating strategy for parallel hybrid powertrains*

For the realization of the coupled simulation consisting of the causal control level modeled in Matlab/Simulink (operating strategy, transmission control, driver model and driving cycle) and the non-causal physical level (physical powertrain) modeled in SimulationX, the approaches described in section 4 were used and compared.

5.1 FMI for Co-simulation as stand alone

In this approach, using the FMI (Functional Mock-up Interface, see [5]) the control level is generated as FMU (Functional Mock-up Unit). This can be imported into SimulationX by means of the FMI -import and integrated into existing models (cp. **Figure 9**). The simulation of the whole model is performed in SimulationX.

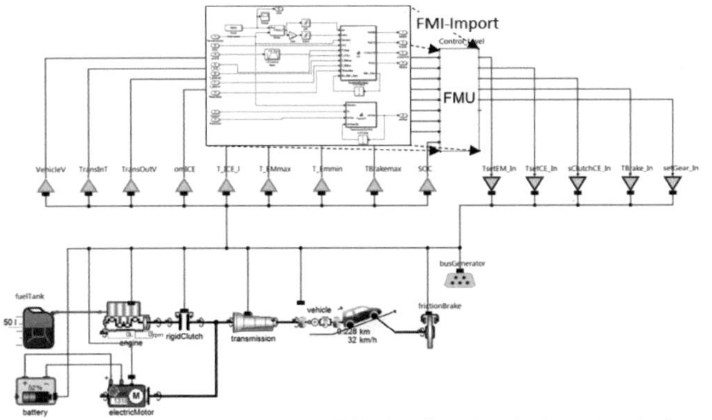

Figure 9: *Co-simulation using FMI for Co-simulation stand alone*

In addition to the model, a user preset fixed-step solver is exported within the FMU. Here, the solver step-size is set before the export. This step-size has a direct impact on the communication step-size between the FMU (control level) and the physical level that can be defined in SimulationX later. The communication step-size *tc* should be smaller than or equal to the controller's fixed step-size. When dimensioning the controller step-size *hc,* the dynamics of the physical level play a major role. To avoid aliasing effects under-sampling, the controller's step-size should be smaller than the smallest time constant of the plant. As a guide, it should be a tenth of the smallest plant's time constant [6]. In our case, a controller step-size of 1 ms has been proven sufficient. Thus, a balance between the required accuracy and the simulation effort could be achieved.

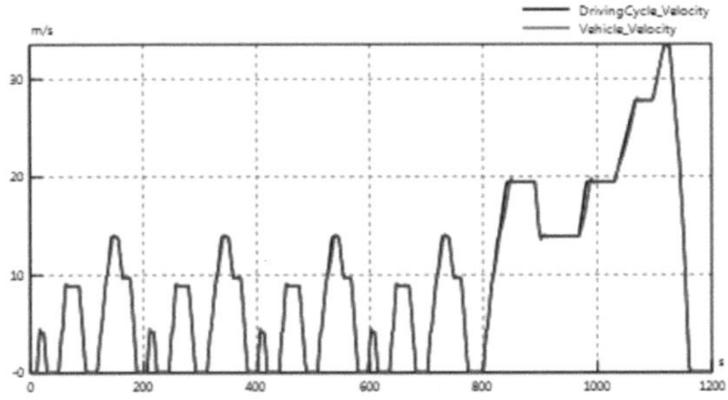

Figure 10: *Comparison of the vehicle velocity with the driving cycle velocity of the holistic modeling and simulation in SimulationX*

To evaluate the simulation approach, the simulation results of the vehicle model by means of the FMI solution were compared with those of the vehicle model generated solely in SimulationX. In both simulation scenarios, the vehicle was stimulated by means of the NEDC cycle. The following figures show the speed profile of the vehicle compared to the cycle velocity. The simulation results of the holistic SimulationX vehicle model (see **Figure 10**) are used as reference for the comparison.
Figure 11 shows the velocity results in the simulation of the powertrain within SimulationX using the FMU for Co-simulation imported from Simulink. The resulting differences between **Figure 10** and **Figure 11**(a) are due to the coupling algorithms and the selected solver step-size of the FMU in Simulink (*hc*=1ms). The communication step-size has no influence here, because it is equal to the solver step-size. In order to increase the simulation accuracy, the solver step-size should be reduced before the export of the FMU. Due to the increased communication step-size (*tc*=100 ms), larger simulation deviations can be seen in the **Figure 11** (b). Whereas the computational effort has been reduced to one tenth of the computational effort when using a *tc* of 1 ms. This can be explained through the lower number of events treated. This can lead to larger value changes (value jumps), such as torque and speed values, which has negative effects on the simulation accuracy and stability.

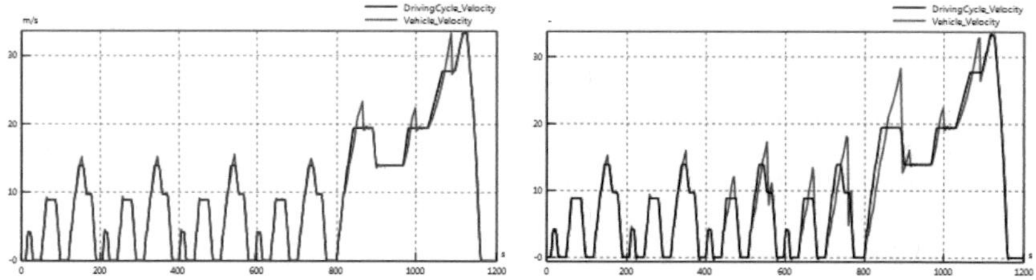

(a): Communication step-size: *tc*=1ms (b): Communication step-size *tc*=100ms

Figure 11: *Comparison of the vehicle velocity with the driving cycle velocity using simulation by means of the FMI for Co-simulation and different tc*

5.2 Multi-tool coupled simulation (Co-simulation)

The multi-tool Co-simulation offers an alternative to the FMI Co-simulation explained bevore. In comparison to the last approach, no models or sub-models have to be exported and imported, which avoids abstraction approximations of the models. For this purpose, the advantages of each modeling paradigm and each simulation environment are combined via an interface for data exchange. In the presented simulation scenario, the control level is calculated in Simulink and the physical level is simulated in SimulationX (see **Figure 12**). Both applications exchange their simulation results at specified time points. To avoid synchronization errors and thus calculation deviation, SimulationX runs as master simulator and Simulink as a slave simulator.
To exchange data between the two simulators, each sub-model (controller and plant) was extended by a coupling element (see **Figure 12**). For each sub-model, the number of inputs and outputs (data to be exchanged), the working mode (slave, master) and the communication step-size (*tc*) can be set through these elements. The communication step-size defines the time between two data exchange instants (communications) between the two simulators. Intermediately both simulators use their pro-

prietary computation step-sizes and an extrapolation of the signal received from the other simulator takes place. The communication between the simulators is achieved via the TCP/IP protocol.

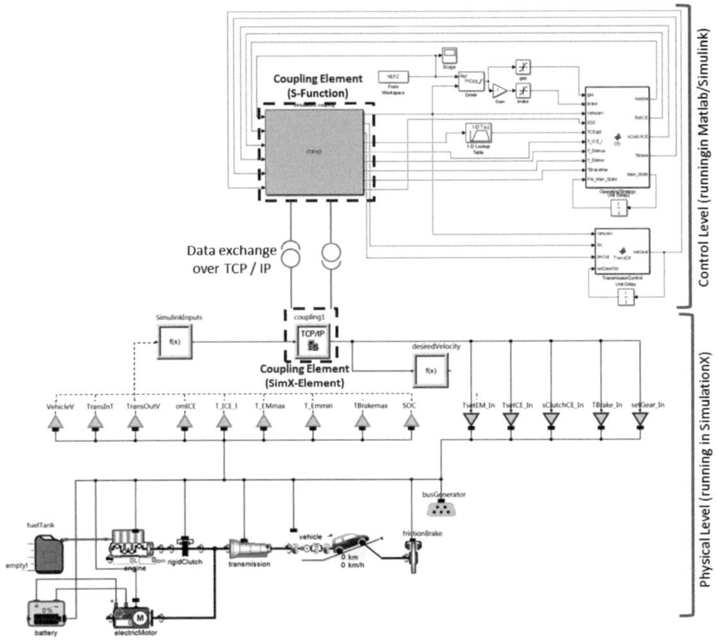

Figure 12: *Co-simulation for the parallel hybrid model using multi-tool Co-simulation between SimulationX and Matlab/Simulink*

In this simulation scenario, the controller model was also calculated in Simulink with a fixed step-size *hc* of 1 ms. The data exchange was effected at a step-size *tc* of 1 ms. The simulation results are similar to those of the FMI Co-simulation. Here, an increased computational effort (ten times) is listed because of the data exchange overhead and results in interpolation/extrapolation [3].

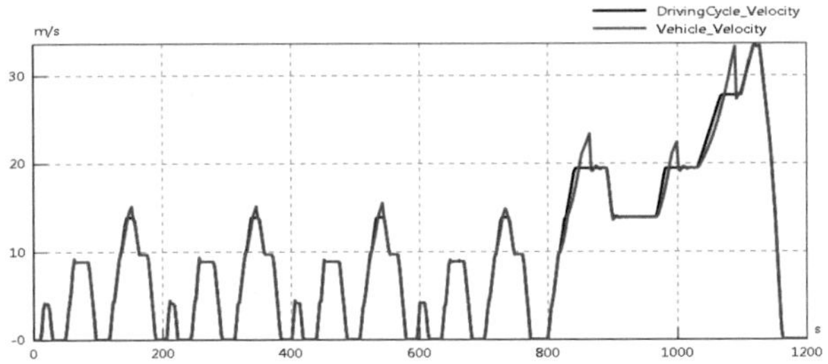

Figure 13: *Comparison of the vehicle velocity with the driving cycle velocity using multi-tool Co-simulation (hc=1ms, tc=1ms)*

6. Conclusion and outlook

The combination of shorter development cycles, ever-growing model complexity, depth of detail, variety and the vast amount of reliable assessments at early stages of the process pose a major challenge in today's everyday work. Beyond modeling and simulating, collection and preparation of input parameters as well as the visualization of the generated results in an interpretable form take up a great deal of the engineer's time. The object oriented, non-causal modeling approach meets the needs of the user, and the intelligent optimization algorithms provide an efficient simulation. Due to the advantages of this modeling paradigm, increasingly heterogeneous domains can be combined and simulated in a single model. The simulation performance reaches the technical limits especially when handling heterogeneous rigid models. The implementation of a distributed simulation using the Co-simulation approaches described above provides an excellent way to overcome this challenge. For this purpose, the heterogeneous overall model is split into several weakly coupled models that can be assigned to one or more modeling and simulation environments. The control loop described in our case represents a weak coupling between the control level and the physical level, since only signals (manipulated variables, command variables) are exchanged between the two units. Depending on the needs and available resources, both units can be distributed to two coupled simulators (Co-simulation). A distinction between the FMI and the multi-tool Co-simulation can be made. When using the first approach, the sub models are exported as FMUs with its own solver and they are calculated in one environment (SimulationX). The second approach allows for coupling different tools (SimulationX and Matlab/Simulink). Both approaches have their advantages and disadvantages which strongly depend on the realization concept and are reflected on the simulation accuracy and performance. The coupling strength (algebraic loops, closed loop controller) between the sub models is significant for the stability of the numerical simulation. The stronger the coupling, the more time-consuming the simulation becomes. The identification of weakly coupled sub-models requires a big effort and profound knowledge of the model especially when handling non-causal physical models, since such models are coupled by means of flow and potential variables and the solving direction is not always clear. Approaches and methods for handling strongly coupled sub-models by means of a Co-simulation are still to be developed.

Literature

[1] Markus Stiegeler. Entwurf einer vorausschauenden Betriebsstrategie für parallele hybride Antriebsstränge. Dissertation, Fakultät für Ingenieurwissenschaften und Informatik, Universität Ulm, 2008.

[2] Stefan Mohr. Entwicklung und Umsetzung eines modellbasierten Vorgehens zur Bewertung hybrider Fahrzeugkonzepte. Studienarbeit, Institut für Automobiltechnik Dresden, Technische Universität Dresden, 2011.

[3] Torsten Blochwitz. Tutorial Model Exchange. 14th ITI Symposium, 2012

[4] User Manual SimulationX, ITI GmbH, 2013

[5] www.fmi-standard.org/

[6] Lars Völker: Untersuchung des Kommunikationsintervalls bei der gekoppelten Simulation, PhD Thesis, Karlsruhe 2010.

13 Requirements to Electrics/Electronics Architectures due to Dual Voltage Power Supplies with 48 Volt in Vehicles

Dipl.-Ing. **Ottmar Sirch**, Dr.-Ing. **Georg Immel**, Dipl.-Ing. **Joachim Fröschl**, Dipl.-Ing. **Alexander Mai**, Dr. rer. nat. **Hartmut Pröbstle**, BMW Group, Munich, Germany

Abstract

The continously increasing requirements for vehicle power supply systems in conventional vehicles with combustion engines have allowed the introduction of 48 Volt as an additional vehicle power supply in the near future. The issue was taken up by Audi, BMW, Daimler, Porsche and Volkswagen at an very early stage and a first specification for the second voltage level with 48 Volt was developed as a common draft LV148. Other car manufacturers deal currently focusing on new system concepts based on dual-voltage vehicle power supply systems with 12 Volt and 48 Volt.

The design and characteristics of such systems and their architectures depend primarily on the CO2 targets and require for the concepts of the dual-voltage vehicle power supply systems a balance between efficiency, weight, package and costs. In addition, clearance for functional increase and innovations should be created that increase customer benefits of the system and thus the vehicle significantly. Beside the new functional requirements and the resulting architecture and dimensioning tasks there is a variety of technical questions in electrics/electronics, which have to be answered very early in the predevelopment phase: electromagnetic compatibility, arcing, short circuits between the voltage levels and the appearance of cross-linking in mixed-powered systems may be mentioned here as examples. The challenge to the management of electrical energy and power consists on the one hand in the management of the sub-system networks with 12 Volt and 48 Volt, and on the other hand in the coordination of the sub-system networks with predictive aspects of using environmental coupling. In this paper, the requirements for the dual voltage vehicle power supply system, the new challenges and the corresponding solution space are presented. Technical details will be explicated by means of case studies.

Kurzfassung

Die ständig steigenden Anforderungen an die elektrische Energieversorgung in konventionellen Kraftfahrzeugen haben die Einführung von 48 Volt als zusätzliche Versorgungsspannung im Kraftfahrzeug in absehbare zeitliche Nähe rücken lassen. Bei Audi, BMW, Daimler, Porsche und Volkswagen wurde die Thematik frühzeitig aufgegriffen und eine erste Spezifikation für eine zweite Spannungsebene mit 48 Volt in Form der Liefervorschrift LV148 erarbeitet. Auch andere Automobilhersteller beschäftigen sich derzeit intensiv mit neuen Systemkonzepten auf Basis von Zwei-Spannungs-Bordnetzen mit 12 Volt und 48 Volt.

Die Gestaltung und Ausprägung solcher Systeme und deren Architekturen hängen in erster Linie mit den CO2-Zielsetzungen zusammen und erfordern von den Lösungskonzepten für das Zwei-Spannungs-Bordnetz ein ausgewogenes Verhältnis zwischen Effizienz, Gewicht, Bauraum und Kosten. Zusätzlich soll Freiraum für funktionale Mehrungen und Innovationen geschaffen werden, die den Kundennutzen des Systems und damit des Fahrzeugs deutlich steigern. Neben neuen funktionalen Anforderungen und den daraus resultierenden Architektur- und Dimensionierungsaufgaben ergibt sich eine Vielzahl an technischen Fragestellungen in der Elektrik/Elektronik, die frühzeitig im Rahmen der Vorentwicklungsphase zu beantworten sind: Elektromagnetische Verträglichkeit, Lichtbögen, Kurzschlüsse zwischen den Spannungsebenen und die Darstellung der Vernetzung in gemischt versorgten Systemen seien an dieser Stelle als Beispiele genannt. Die Herausforderung an das Management von elektrischer Energie und Leistung besteht einerseits im Management der Teilbordnetze mit 12 Volt und 48 Volt und andererseits in der Koordination der Teilbordnetze unter prädiktiven Aspekten mittels einer Umweltkopplung. In diesem Beitrag werden die Anforderungen an das 2-Spannungs-Bordnetz, die neuen Herausforderungen und der korrespondierende Lösungsraum vorgestellt. Anhand von Fallbeispielen werden die technischen Details erläutert.

1. Einleitung

Seit Mitte der 90er Jahre hat sich der Energie- und Leistungsbedarf im Bordnetz von konventionellen Kraftfahrzeugen mit Verbrennungsmotor überproportional erhöht. Neue E/E-Kundenfunktionalitäten führen zusammen mit fortschreitender Elektrifizierung von Antriebsfunktionen und Nebenaggregaten - zur Erreichung der angespannten CO2-Ziele - das elektrische Energiebordnetz im Kraftfahrzeug an die Grenzen seiner Belastbarkeit. Besonders hervorzuheben sind an dieser Stelle die Auto-Start-Stopp-Funktion, die Bremsenergierückgewinnung und die Elektrifizierung mechanisch betriebener Systeme, wie z.B. die elektrische Lenkkraftunterstützung. Neben dem zunehmenden Energiebedarf einzelner Funktionen stellt vor allem auch die Überlagerung mehrerer Hochstromverbraucher eine besondere Herausforderung an die Leistungsfähigkeit zukünftiger Bordnetzarchitekturen dar.

Auch wenn absehbar ist, dass in Zukunft Mobilität über elektrifizierte Antriebsstränge definiert wird [5], bleiben effiziente verbrennungsmotorische Antriebe weiterhin eine unverzichtbare Option. Die Effizienz von Verbrennungsmotoren lässt sich u.a. durch Reduktion der Verlustleistung seiner Aggregate, wie z.B. der Generatorverluste massiv steigern. Ermöglicht wird dies z.B. durch „starke“ Bordnetze, die einen generatorischen Betrieb vorwiegend in den Brems- oder Verzögerungsphasen des Fahrzeugs benötigen. Dies bedingt wiederum zyklenfeste Speicher und leistungsfähige Generatoren, die das vorhandene Rekuperationspotenzial effizient ausnutzen. Es ist hier wichtig zu erwähnen, dass sich der aktuell oftmals propagierte „Segelbetrieb bei abgestelltem Motor“ unter der Randbedingung konventioneller motorgebundener Generatoren nachteilig auf das Rekuperationspotenzial auswirkt. Ein abgekoppelter Generator kann in einem solchen Fall die Bremsenergie nicht in wertvolle Rekuperationsenergie umwandeln. „Segeln“ bei abgeschalteten Motor macht daher aus Energiebordnetzsicht nur dann Sinn, wenn gleichzeitig der „generatorische Druck“ in den Verzögerungsphasen aufrechterhalten bleibt, was neue An-triebsarchitekturen mit z.B. getriebegebundenen Generatoren erfordert. Die Energiemenge, die bei einem Bremsvorgang rekuperiert werden kann, hängt - bei vergleichbarer Antriebsarchitektur - vom Fahrzeuggewicht und der Fahrzeuggeschwindigkeit ab. Bei großen Fahrzeugen ist die Verzögerungsleistung und damit das theoretische Rekuperationspotenzial höher. In der Realität wird dieses Potenzial durch die Leistungsfähigkeit der verbauten Systemkompenten wie Generator und Energiespeicher begrenzt. Hier bietet z.B. die Anhebung der Systemspannung von 12 Volt auf 48 Volt Potenziale. Erste Untersuchungen haben gezeigt, dass bei kleinen Fahrzeugen mit „starken" 12-Volt-Bordnetzen bereits eine Kompensation des Bordnetzeigenbedarfs über Rekuperation möglich ist. Bei großen Fahrzeugen mit hoher Ausstattung und typischerweise hohem Bordnetzbedarf bieten 48-Volt-Rekuperations-architekturen die Möglichkeit einer 100%-igen Eigenbedarfsdeckung.

Mit einer 48-Volt-Elektromaschine eröffnen sich zudem weitere Möglichkeiten zur CO2-Einsparung, da sie sowohl im generatorischen als auch im motorischen Betrieb den Lastpunkt des Verbrennungsmotors in einen günstigeren Betriebspunkt verschieben kann. Die neuen Anforderungen nach höherer elektrischer Leistung werden zukünftige Energiebordnetz-Architekturen hinsichtlich Topologie, Gesamtstromverbrauch und funktionaler Ausprägung maßgeblich beeinflussen und verändern. Als Beispiel hierfür zeigt Bild 1 ein Zwei-Spannungs-Bordnetz klassischer Ausprägung. Durch die Entkoppelung der Bordnetzzweige können die von Hochleistungsverbrauchern verursachten Wechselwirkungen zwischen den verschiedenen Bordnetzzweigen verringert oder sogar verhindert werden.

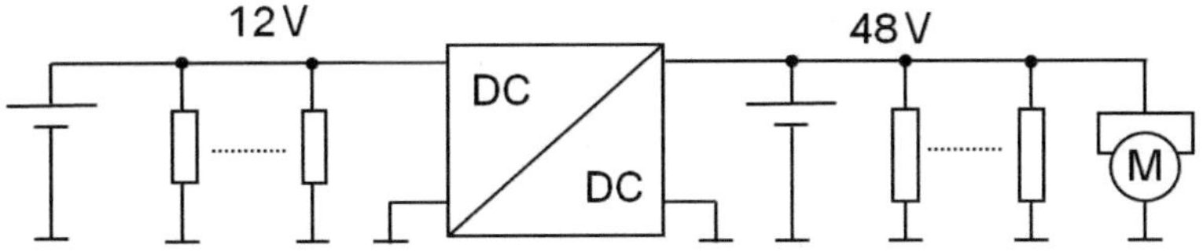

Bild 1: *Schematische Darstellung eines Zwei-Spannungs-Bordnetzes*

Weitere Ausprägungen des Zwei-Spannungs-Bordnetzes sind denkbar und werden in Kapitel 3 behandelt.

2. Änderung der Anforderungen

Bis 2020 ist eine signifikante Veränderung der Rahmenbedingungen für die Bewertung des CO2-Ausstoßes von Kraftfahrzeugen in Europa zu erwarten. Der Normverbrauchszyklus soll vom Neuen Europäischen Fahrzyklus NEDC auf den Worldwide Harmonized Light Vehicle Test Procedure WLTP umgestellt werden. Die unterschiedlichen Geschwindigkeitsprofile von NEDC und WLTP sind in Bild 2 dargestellt.

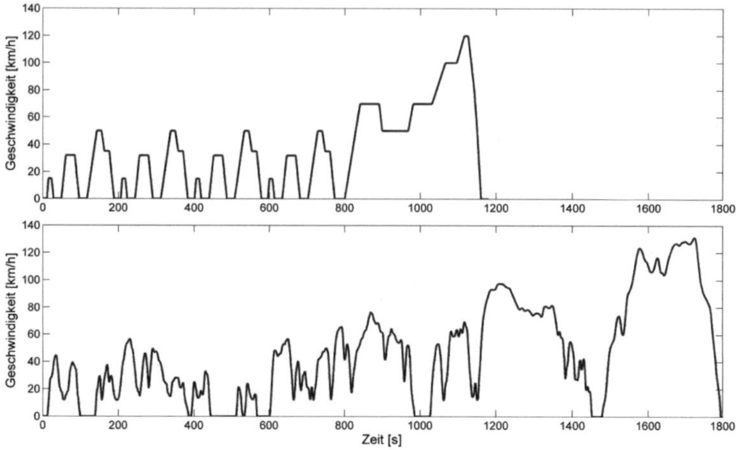

Bild 2: *Geschwindigkeitsprofile NEDC (oben) im Vergleich zu WLTP (unten)*

Während der NEDC über einen hohen Anteil an Stopp-Phasen verfügt, ist der zukünftig von der Europäischen Union präferierte WLTP durch erheblich weniger Stopp-Phasen gekennzeichnet. Beim WLTP werden im Gegensatz zum NEDC Sonderausstattungen, allerdings keine Einschaltprofile von Komfortfunktionen berücksichtigt. Eine deutliche Erhöhung der Verzögerungsanteile von 31% auf 43% begünstigt die Gewinnung von elektrischer Energie durch Rekuperation.

Vergleich hinsichtlich der maßgeblichen Parameter zwischen NEDC und WLTP ist in Tabelle 1 dargestellt.

Parameter	**NEDC**	**WLTP**
Wegstrecke [km]	11,013	23,141
Dauer [s]	1180	1800
Mittl. Geschwindigkeit [km/h]	33,6	46,3
Max. Geschwindigkeit [km/h]	120	131
Max. Beschleunigung]m/s2]	1,04	1,88
Max. Verzögerung [m/s2]	-1,39	-1,52
Standzeiten [s]	280	227
Anteil Standzeiten [%]	23,7	12,6
Anteil Verzögerung [%]	31,0	43,1

Tabelle 1: *Vergleich zwischen NEDC und WLTP*

3. Architekturen für Zwei-Spannungs-Bordnetze

Das elektrische Energiebordnetz im Fahrzeug hat in den letzten 10 Jahren einen erheblichen Anstieg der Anforderungen, jedoch keine grundlegende Veränderung in Architektur und Topologie erfahren. Die Integration neuer Funktionen wurde dabei stets über die Dimensionierungen von Klauenpol-Generator, Bleibatterie, Kabelbaum und Stromverteiler im „standardisierten“ 12-Volt-Bordnetz (Bild 3) umgesetzt, was auf einen Mehrbedarf an

- elektrischer Energie durch stationäre Stromverbraucher (z.B. Heizsysteme, Motorlüfter),
- elektrischer Leistung durch transiente Stromverbraucher (z.B. Lenkkraftunterstützung, Fahrwerksregelsysteme, Warmstart) und
- Speicherzyklisierung durch Energiedurchsatz (z.B. Bremsenergierückgewinnung, Vor-/Nachlauf, Versorgung in der Motorstopphase)

zurückzuführen ist.

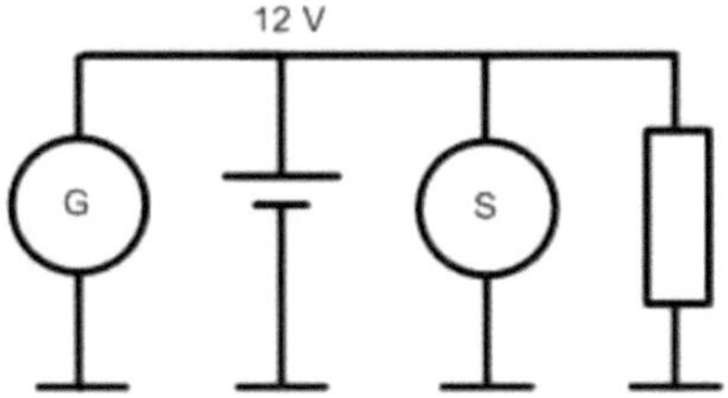

Bild 3: *Schematische Darstellung der heutigen Energiebordnetzarchitektur*

Durch ständige Optimierungsschritte und Verbesserungen sowohl bei der Generierung als auch bei der Speicherung von elektrischer Energie kann die Leistungsfähigkeit dieser Bordnetzarchitektur weiter gesteigert werden. Dies geschieht auch durch die Einführung von Effizienzmaßnahmen, die den Stromverbrauch von Funktionen, Systemen und Aggregaten reduzieren, und somit indirekt die Energie- und Leistungsverfügbarkeit erhöhen. Ein anstehender Architektursprung, wie z.B. die Einführung eines Zwei-Spannungs-Bordnetzes, kann unter Umständen dadurch hinausgezögert werden. Durch die sukzessive Einführung von Maßnahmen zur Reduzierung des elektrischen Verbrauchs bestehender Funktionen [1], wie in Bild 4 dargestellt, kann Freiraum zur Einführung neuer Funktionen geschaffen werden. Je höher die Reduzierung zu einem bestimmten Zeitpunkt ist, desto höher ist das Potenzial für die Einführung neuer Funktionen. Wird allerdings durch den hohen Leistungsbedarf einer neuen Funktion die Schwelle zu einer leistungsfähigeren Architektur überschritten, so kann der Architektursprung nicht länger vermieden werden. Da Architektursprünge mit erheblichen Kosten verbunden sind, sollten sie über kundenwerte Funktionsmehrungen legitimiert werden. Aufbauend auf die einleitend diskutierte und in Bild 3 dargestellte klassische Basisarchitektur bieten sich verschiedene Architekturvarianten in Abhängigkeit von den erweiterten Anforderungen an. Sämtliche Varianten an elektrischen Energieversorgungsarchitekturen lassen sich auf

- verbraucherseitige,
- generatorseitige und
- speicherseitige

Erweiterungen der in Bild 3 gezeigten Basisarchitektur zurückführen. Eine ausführliche Beschreibung hierzu findet sich in [2]. Bei Überschreitung der stationären und dynamischen Leistungsgrenzen des 12-Volt-Bordnetzes bietet sich die Integration einer oder mehrerer Funktionen in Form einer verbraucherseitigen Erweiterung auf 48 Volt an. Bild 4 zeigt das prinzipielle Schaltbild einer verbraucherseitigen 48-Volt-Erweiterung, die vor allem zur Stützung von Hoch-leistungsverbrauchern auf ein entkoppeltes Teilbordnetz zurückgreift. Die verbraucherseitige Erweiterung ist jeweils mit oder ohne Energiespeicher ausführbar, je nach Energie- oder Leistungsbedarf der versorgten Funktionen. Verbraucherseitige Erweiterungen ohne 48-Volt-Speicher sind für stationäre 48-Volt-Verbraucher, die eine höhere Betriebsspannung benötigen (z.B. auf Beschichtung basierende Frontscheibenheizung) geeignet, sofern die Leistungsabnahme die Belastung der 12-V-Seite nicht überschreitet.

Für dynamische Kurzzeitverbraucher wie z.B. Lenk- oder Wankstabilisierungssysteme werden vorzugsweise Doppelschichtkondensatoren mit hoher Leistungsfähigkeit und Zyklenfestigkeit eingesetzt.

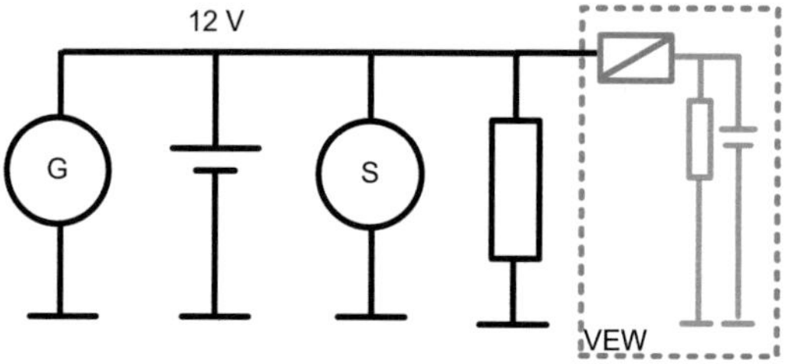

Bild 4: *Architekturprinzip der verbraucherseitigen Erweiterung VEW*

Signifikante energetische Mehrbedarfe sollten vorzugsweise über Speicherarten wie z.B. Lithium-Ionen-Batterien versorgt werden, um eine bedarfsorientierte Versorgung mit minimaler Belastung des Basisbordnetzes zu gewährleisten (z.B. im Start-Stopp-Betrieb).

Stationär erhöhte Energiebedarfe (z.B. Klimafunktionen) oder Rekuperationsanforderungen mit Spitzenleistungen über 5 kW sollten vorzugsweise über die in Bild 5 gezeigte generatorseitige Erweiterungen bedient werden. In dieser Leistungsklasse bietet sich zu-dem ein elektromotorischer Betrieb an, der über die Boost-Funktion $CO2$ -günstige Lastpunktverschiebungen des Verbrennungsmotors ermöglicht.

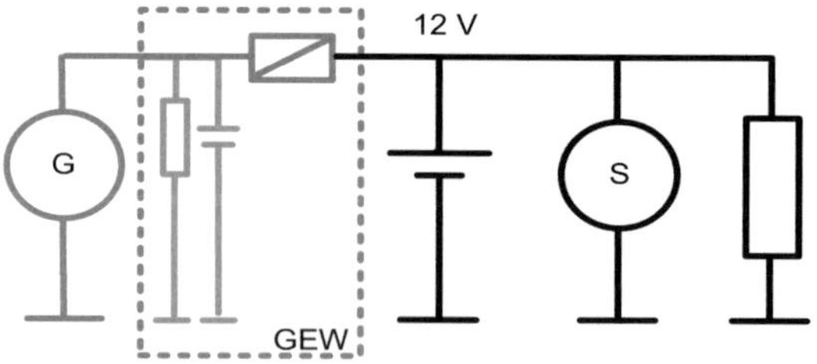

Bild 5: *Architekturprinzip der generatorseitigen Erweiterung GEW*

4. Spezifikation des Spannungsbereichs

Der Spannungsbereich für 48 Volt ist in der LV148 spezifiziert. Bild 7 zeigt die detaillierte Aufteilung des statischen Spannungsbereichs, der für jede 48-Volt-Komponente gilt. In der LV148 sind außerdem die Komponententests detailliert beschrieben.

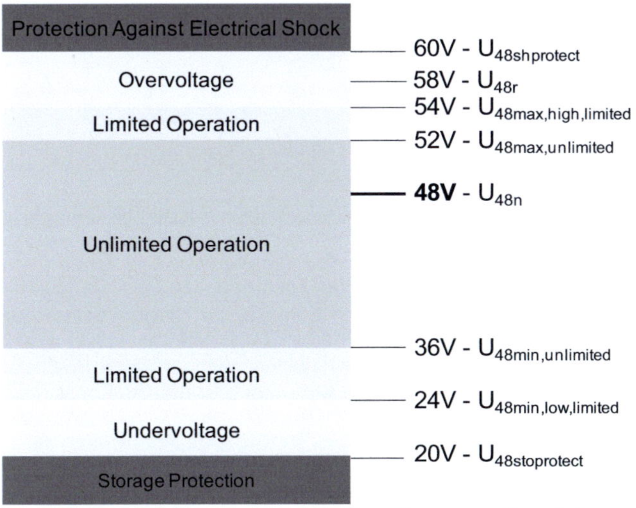

Bild 6: *Definitionen der statischen Spannungsbereiche*

5. Herausforderungen

Die Einführung einer zweiten Spannungsebene deutlich größer als 12 Volt bringt eine Vielzahl an Herausforderungen an die elektrischen Komponenten und deren Integration mit sich. Einige dieser Anforderungen, die sich als besonders wichtig und systemrelevant herauskristallisiert haben, sollen im Folgenden näher erläutert werden.

5.1 Elektrisches Energie- und Leistungsmanagement

Die Herausforderung an das flexible Management von elektrischer Energie und Leistung besteht einerseits im Management der Teilbordnetze mit 12 Volt und 48 Volt und andererseits in der Koordination der Teilbordnetze unter prädiktiven Aspekten mittels einer Umweltkopplung, wie schematisch in Bild 7 dargestellt.

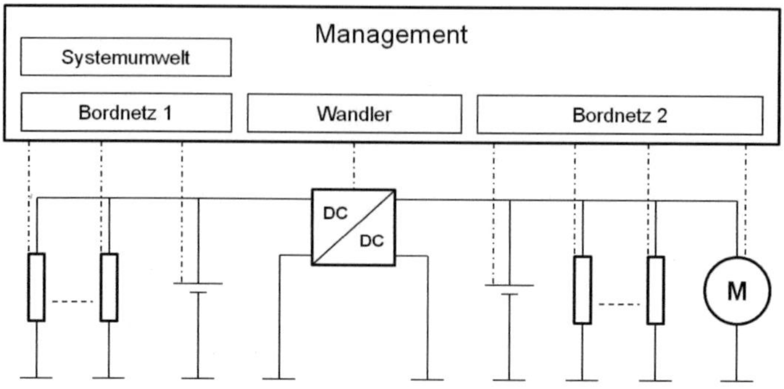

Bild 7: *Prinzip des elektrischen Energie- und Leistungsmanagements*

Bei einem Ein-Spannungs-Bordnetz besteht die Aufgabe des flexiblen Energie- und Leistungsmanagements darin, sowohl die Zustände von Energiehaushalt, -reserve, und -tendenz als auch die Zustände von Leistungshaushalt, -reserve, und -tendenz in einem stabilen Arbeitsbereich zu halten [11]. Werden zwei Ein-Spannungs-Bordnetze gekoppelt, so gelten diese Aufgaben gleichermaßen für beide Teilbordnetze. Eine ergänzende Aufgabe ergibt sich jedoch aus der Koordination der Teilbordnetze [4]. Diese besteht darin, die beiden Bordnetze in einem ausgeglichenen Zustand zueinander zu halten. Der Ausgleich muss gleichermaßen für Energie und Leistung unter Berücksichtigung von Stabilität und Effizienz erfolgen. Unter Berücksichtigung von Informationen aus der Systemumwelt, wie beispielsweise einer Navigationskopplung, kann dieses Gleichgewicht situationsgerecht modifiziert werden [12]. Es muss aus Sicht der koordinierenden Betriebsstrategie entschieden werden, ob beispielsweise für eine Rekuperationsphase zunächst prädiktiv eine Konditionierung, d.h. Absenkung des Speicherladezustands , für ein oder beide Teilbordnetze erfolgen sollte, um anschließend den einen oder beide elektrischen Speicher rekuperativ zu befüllen. Dabei muss für beide Teilbordnetze die Stabilität der Betriebsspannung zu jedem Zeitpunkt gewährleistet sein. Das Managementsystem muss weiterhin die oben gezeigten Architekturvarianten optimal bedienen können. Dabei ist eine standardisierte Anbindung der Komponenten [12] einerseits von wirtschaftlicher Bedeutung und andererseits von technischem Vorteil hinsichtlich der Austauschbarkeit und Migrationsfähigkeit innerhalb des architektonischen Lösungsraumes. Die Betriebsstrategie muss dabei den hinsichtlich Effizienz optimalen Einsatz der verwendeten elektrischen Komponenten als Gesamtsystem mit dem Ziel der bestmöglichen Stabilität der einzelnen Systemspannungen sicherstellen. Ebenso ist der Einsatz des DC/DC-Wandlers innerhalb seiner Leistungsgrenzen aber auch möglichst lange im Bereich seiner maximalen Effizienz sicherzustellen. Hierzu bieten sich die Möglichkeiten der Erhöhung des Angebots und der Degradation der Verbraucher in beiden Teilnetzen an. Eine Prädiktion der Erzeugungs- und Verbrauchssituation mittels Umweltkopplung unterstützt dabei die Entscheidungen der Betriebsstrategie hinsichtlich der Steuerung des Energie- und Leistungstransfers des Wandlers. Ein weiterer Aspekt ist die Behandlung von Fehlern.

5.2 Lichtbögen

Bereits bei Bordnetzspannungen über 12 Volt können stabile Störlichtbögen entstehen. Daher muss dieses Phänomen bei allen Spannungsklassen unter 60 V genau analysiert werden. Sowohl in Personenkraftwagen mit 12 Volt als auch in Nutzkraftfahrzeugen mit 24 Volt tritt das Phänomen heute sporadisch auf und wird akzeptiert. Mit der Einführung von 48 Volt kommt die bereits bei 42 Volt geführte Diskussion erneut auf. Die physikalischen Zusammenhänge in Abhängigkeit von verwendeten Leitungsmaterialien und anliegenden Spannungen sowie weiteren Parametern sind in [6] ausführlich erläutert. Für die weiteren Betrachtungen wird zwischen parallelen und seriellen Lichtbögen unterschieden.

5.2.1 Parallele Lichtbögen

Parallele Lichtbögen können durch Berührung zwischen einer spannungsführenden 48-Volt-Leitung und Masse hervorgerufen werden. Fehler dieser Art sind konstruktiv, z.B. durch Leitungsverlegung und Auswahl geeigneter Isolation, zu vermeiden.

5.2.2 Serielle Lichtbögen

Das Trennen unter Last, hervorgerufen durch Abziehen oder Aufstecken des Steckers oder durch Leitungsabriss, kann zur Bildung eines Lichtbogens führen. Einflussfaktoren sind hier Spannung, Stromstärke und Trenngeschwindigkeit. Da der Lichtbogen-Mindeststrom sehr klein ist, kann sich auch bei Spannungsquellen geringer Leistung bei Überschreiten der Mindestspannung (je nach Leitermaterial, z.B. Kupfer 13 V) ein stabiler Lichtbogen entwickeln. Der minimal notwendige Haltestrom für Lichtbögen liegt bei 0,4 A [6]. Durch den Einsatz eines Kondensators am 48-Volt-Eingang eines Verbrauchers kann die Bildung von seriellen Lichtbögen unter normalen Bedingungen verhindert werden. Dazu muss der Kondensator so dimensioniert werden, dass bei einer Trennung von der elektrischen Versorgung die Spannungsdifferenz an der Trennstelle 12 Volt nicht überschreitet. Dies gilt innerhalb der Zeitdauer, die eine typische Trennung benötigt, um einen ausreichenden Abstand von 0,02 mm zu erreichen. Die Dimensionierung des Kondensators hängt von der Größe des Laststroms ab und ist über folgenden Zusammenhang zu ermitteln:

$$\tau = C \times R \times ln\,\frac{U0}{U1}$$

τ: *Zeitkonstante*

C: *Kapazität des ZK – Kondensators*

R: *resultierender Widerstand der Last*

$U0$: Startwert Eingangsspannung

$U1$: Endwert Eingangsspannung

Ist die sich einstellende Distanz in der Trennstelle innerhalb der Zeitdauer größer oder gleich 0,02 mm, so tritt anschließend bei Zunahme der Spannungsdifferenz über der Trennstelle kein Lichtbogen mehr auf, da die kritische Feldstärke für die Ionisation von Luft (30 kV/cm) nicht mehr erreicht werden kann.

5.3 Kurzschlüsse

In einem Zwei-Spannungs-Bordnetz gelten die gleichen Anforderungen hinsichtlich Kurzschlüssen wie im 12-Volt-Bordnetz. Kurzschlüsse nach Masse von 48 Volt nach Masse müssen genauso sicher erkannt und abgeschaltet werden. Der Kurzschluss von 48 Volt nach 12 Volt ist ein neuer Fehlerfall, der zu einer Zerstörung von 12-Volt-Komponenten führen kann. Er ist durch geeignete konstruktive Maßnahmen, z.B. Leitungsverlegung und Isolation, auszuschließen.

5.4 Kommunikation in gemischten Netzwerken

Die Herausforderungen für Kommunikationssysteme in gemischten Netzwerken liegen vor allem in Signalintegrität und Robustheit [9]. Auch wenn in diesen Bordnetzen die elektrischen Massen beider Spannungsebenen elektrisch miteinander verbunden sind, stellt sich die Frage, welche Auswirkungen dies auf die Fahrzeugkommunikationsnetzwerke hat, die diese beiden Spannungswelten logisch verbinden. Im Beitrag [9] werden im Detail die zu erwartenden Systemkonstellationen bei der Verwendung verschiedener Bussysteme erörtert. Gezeigt werden dort denkbare Systemzustände und deren Auswirkungen. In Kommunikationssystemen, basierend auf CAN oder FlexRay, die eine differentielle Übertragung nutzen, haben die Bus-Transceiver trotzdem einen gemeinsamen Massebezug. Im Falle eines Masseverlusts eines 48-Volt-Busteilnehmers treten bei nur 2 Busteilnehmern, die identische Bustransceiver einsetzen, folgende Spannungsdifferenzen an den Transceiver-Eingängen auf: Am mit Masse verbundenen Transceiver fallen UBat48/2 ab, und am nicht mit Masse verbundenen Transceiver liegt eine Spannungsdifferenz von -UBat48/2 an. Je mehr Transceiver im gleichen Netzwerk noch eine Masse haben, desto größer wird die Spannungsdifferenz zum Transceiver ohne Masse bis nahe an -UBat48. Wenn der Masseabriss während der aktiven Kommunikation stattfindet, kann auch der masselose Transceiver noch eine gewisse Zeit weiter senden und Pegel bis nahe an UBat48 auf den Bus einprägen. Daraus leiten sich folgende Anforderungen an die verwendeten Transceiver ab [9]:

- In 48-Volt-Komponenten eingesetzte Transceiver müssen robust für negative Spannungen bis +/-58 V sein.
- In 12-Volt-Komponenten eingesetzte Transceiver müssen eine Spannungsfestigkeit bis 58 V aufweisen.
- Harte Kurzschlüsse zwischen 48 Volt und einer der Busleitungen müssen durch konstruktive Maßnahmen verhindert werden.
- In Komponenten mit beiden Versorgungsspannungen muss die Anforderung „Querstrom kleiner 1 µA“ (gem. LV 148 Test E48-20) erfüllt werden.

5.5 Power-Up und Power-Down

Für die Inbetriebnahme und die Beendigung des Betriebs von 48-Volt-Systemen sind folgende Randbedingungen zu beachten.

Inbetriebnahme: Der 48-Volt-Speicher ist im abgestellten Zustand einpolig abgeschaltet. Vor dem Schließen des Schalters müssen die Zwischenkreise der 48-Volt-

Komponenten vorgeladen werden, damit wegen der entladenen Kapazitäten keine hohen Ladeströme fließen. Die Inbetriebnahme muss innerhalb der Aufstartsequenz des Gesamtfahrzeugs so frühzeitig erfolgen, dass es zu keinem Fehlverhalten in den Kommunikationsabläufen der Netzwerke kommt.

Beendigung: Nach einer in Abhängigkeit von Nachlauffunktionen zu definierenden Zeit ist der Schalter des 48-Volt-Speichers zu öffnen und die Zwischenkreise sind auf einen definierten Spannungswert zu entladen. Vorher müssen die relevanten Kommunikationsabläufe definiert beendet werden.

5.6 Elektromagnetische Verträglichkeit

Die elektromagnetische Verträglichkeit (EMV) spielt bei der Zulassung neuer Fahrzeuge eine wichtige Rolle. Elektromagnetische Störungen, die durch das 48VoltSystem erzeugt werden, müssen den gleichen Richtlinien und Grenzwerten genügen, die für herkömmliche Fahrzeugkonzepte mit 12 Volt gelten [8]. Diese Anforderungen sind bereits beim Entwurf der elektrischen Architektur des Fahrzeugs und der Komponenten zu berücksichtigen, um das System EMV-technisch ordnungsgemäß zu gestalten. Die hohen Spannungs- und Stromgradienten du/dt und di/dt erfordern eine entsprechende Berücksichtigung bei der Auswahl der Topologie und geeignete Filtermaßnahmen in den Komponenten. Hier bedarf es bereits in frühester Phase einer intensiven Zusammenarbeit zwischen Fahrzeughersteller und Komponentenlieferanten, um ein Optimum zwischen Filtermaßnahmen und Bordnetzgestaltung zu finden.

5.7 Umgang mit Spannungen < 60 Volt

Der Spannungsbereich wurde gezielt so spezifiziert, dass ein Überschreiten der Berührungsschutzgrenze von 60 Volt_DC ausgeschlossen ist. In der LV148 ist folgende Prämisse postuliert: Es ist kein Berührungsschutz für Gleichspannungen ≤ 60 V erforderlich. Dies gilt für Gleich-spannungen bis zu einer Bord-netzwelligkeit von maximal 10% RMS. Bei Wechselspannung bis Ueff ≤ 30 V ist kein Berührungsschutz notwendig (ISO 6469-3).

Ferner gilt die ECE-R100. Ein entsprechender Änderungsantrag zur ECE-R100 wurde in die entsprechenden Gremien eingesteuert und wurde Ende 2013 bestätigt.

6. Chancen durch die Einführung der neuen Spannungsebene

Neben der höheren Leistungsfähigkeit und den damit verbundenen funktionalen Erweiterungsmöglichkeiten bietet das Zwei-Spannungs-Bordnetz mit 48 Volt weitere Potentiale hinsichtlich dem Gewicht des Kabelbaums und der Effizienz der elektrischen Subsysteme.

6.1 Gewichtsreduzierung des Kabelbaums

Bei der Auslegung physischer 12-Volt-Bordnetze geht aus Gewichtsgründen der Trend aktuell von Kupfer zu Aluminium. Nachteilig wirken sich hier in Einzelfällen der höhere Leitungsquerschnitt bei gleichem Strom und die schwerer beherrschbare Verbindungstechnik zu kupferbasierten Teilkomponenten aus. Das 48-Volt-Leistungsbordnetz kann mit Kupfer oder Aluminium als Leitungsmaterial ausgelegt werden. Das Leitungsgewicht wird dort unabhängig vom Leitungsmaterial durch den geringeren Leiterquerschnitt, verursacht durch den geringeren Strom, deutlich reduziert, und bei nicht vollständiger Ausschöpfung der Querschnittsreduzierung können außerdem die Leitungsverluste verringert werden. Das Optimum aus Gewicht und Verlusten ist im Einzelfall zu ermitteln und in Abhängigkeit von Bauraumvorgaben zu betrachten.

6.2 Effizienzsteigerung durch Verlagerung von Verbrauchern

Bei generatorseitigen Erweiterungen ist aus Effizienz- und Dimensionierungsgründen eine Verlagerung von leistungsstarken Verbrauchern auf die 48-Volt-Seite sinnvoll. Damit werden die Wandlungsverluste großer Leistungsflüsse von 48 Volt nach 12 Volt über den DC/DC-Wandler in Höhe von 3% bis 5% vermieden. Ist die Verlagerung von 12-Volt-Verbrauchern auf die 48-Volt-Seite geeignet gewählt, kann die DC/DC-Wandler-Größe bei ausgewogener Konfiguration der Leistungsverbraucher in den verschiedenen Betriebssituationen (z.B. Sommer- vs. Winterbetrieb), deutlich reduziert werden. Für die Verlagerung von existierenden 12-Volt-Hochleistungsverbrauchern eignen sich der elektrische Motorlüfter, dessen stationäre Leistungsaufnahme bis zu 1 kW erreichen kann, elektrische Heizsysteme (z.B. PTC), deren stationäre Leistungsaufnahme heute bereits über 1 kW liegt, und verschiedene elektrische Pumpen. Aus dem Bereich Fahrdynamiksysteme sind z.B. die elektrische Lenkkraftunterstützung und die elektrische Wankstabilisierung zu nennen, die durch ihre hohen transienten Leistungsaufnahmen das heutige 12-Volt-Bordnetz destabilisieren.

7. Ausblick

Mit den genannten Herausforderungen sind wichtige Themen für die Elektrik/Elektronik von Zwei-Spannungs-Bordnetzen mit 48 Volt andiskutiert und erste Lösungsansätze beschrieben. Weitere Themen, die diskutiert werden müssen, sind z.B. Crash-Abschaltung, Zuverlässigkeit von elektrischen Kontakten und Bordnetzstabilität im Falle transienter Hochleistungsverbraucher sowie die Relevanz hinsichtlich funktionaler Sicherheit und der Umgang in Produktion, Montage, Transport und Service.

Langfristig ist zu erörtern, welche Verbraucher letztendlich auf 48 Volt verlagert werden können und welche neuen Funktionen, die heute mit 12 Volt nicht elektrifiziert werden können, dort dargestellt werden.

Literatur

[1] Ottmar Sirch, Georg Immel, Hartmut Pröbstle, Rupert Neudecker, Joachim Fröschl, Zukunft Energiebordnetz - Von der energetischen Optimierung zum neuen Gesamtkonzept, VDI 14. Internationaler Kongress Elektronik im Kraftfahrzeug, Baden-Baden, Oktober 2009.

[2] Hartmut Pröbstle, Rupert Neudecker, Ottmar Sirch, Power Supply in Future Start-Stop-Systems, Energiemanagement und Bordnetze IV, München, März 2011.

[3] T. Dörsam, Daimler AG, Böblingen, D. Groh-mann, Daimler AG, Sindelfingen, S. Kehl, Porsche AG, Weissach, A. Klinkig, Volkswagen AG, Wolfsburg, A. Mai, BMW Group, München, O. Sirch, BMW Group, München, A. Radon, Audi AG, Ingolstadt, J. Winkler, Audi AG, Ingolstadt, Die neue Spannungsebene 48 Volt, VDI 15. Internationaler Kongress Elektronik im Kraftfahrzeug, Baden-Baden, Oktober 2011.

[4] Fröschl, T. Kohler, A. Thanheiser, H.-G. Herzog, Intelligente Bordnetzkopplung am Beispiel Zweispannungsbordnetze mit 12 V und 48 V, Elektrik/Elektronik in Hybrid- und Elektrofahrzeugen und Elektrisches Energiemanagement, Miesbach, April 2012.

[5] Herausforderungen und Potenziale von 48-Volt-Startsystemen, Michael Timmann, Martin Renz, Oliver Vollrath, ATZ, März 2013

[6] Meckler, Störlichtbögen in Automotive-Bordnetzen, Fahrzeugelektronik 6/2012

[7] J. Bast, M. Kilger, W. Galli, A. Eiser, H.-W. Vaßen, I. Kutschera, Die Chancen des Antriebsstrangs durch das 48V-Bordnetz, 34. Internationales Wiener Motorensymposium 2013

[8] Franck Briault, Salah Benhassine, Marco Klingler, Michel Maher, Naguib Rezkalla, EMC Simulation Approach for 48V Systems Integration, eehe 2013

[9] Matthias Muth, Bus Systems within 48V Vehicle Networks – New Challenges for the Physical Layer, eehe 2013

[10] Paul Boucharel, Alexander Nuss, Thomas Knorr, Tobias Galli, Reduction of CO2 in a Low-Voltage Mild Hybrid Vehicle – Conditions, Challenges, and Realization, Electric & Electronic Systems in Hybrid and Electric Vehicles and Electrical Energy Management, Bamberg, April 2013

[11] Joachim Fröschl, Dominik Ilsanker, Hans-Georg Herzog, Investigation of a State Triggered Energy Management System, Electric & Electronic Systems in Hybrid and Electric Vehicles and Electrical Energy Management, Bamberg, April 2013.

[12] Tom P. Kohler, Joachim Fröschl, Hans-Georg Herzog, Systemansatz für ein hierarchisches, umweltgekoppeltes Powermanagement, Elektrik/Elektronik in Hybrid und Elektrofahrzeugen, München, März 2010.

14 Holistic approach for the design of power supply systems in an early development phase for future vehicles

Dipl.-Ing. **Sven Löwenberg**, Prof. Dr.-Ing. **Bernard Bäker**, Technische Universität Dresden, Dresden, Dipl.-Ing. **Victor Schäfer**, Dr.-Ing. **Steffen Kehl**, Dr.-Ing. **Wolfgang Herdeg**, Dr. Ing. h. c. F. Porsche AG, Weissach, Germany

Abstract

In the future, dimensioning of on-board power supply systems for vehicles will play an increasingly significant role. In the early stages of development, the requirements of on-board power supply systems are defined by new functions that reduce fuel consumption, hybridize powertrains and integrate high-current consumers. The integration of multiple voltage levels, wiring, the dimensions of storage systems and energy sources, as well as the migration of loads represent the essential degrees of freedom within the design. Because there are so many degrees of freedom, there are several design options that fulfill these requirements. However, the design should also be optimized in weight and cost efficiency.
This article offers a holistic approach to solving this trade-off.

Kurzfassung

Die Auslegung eines Fahrzeugenergiebordnetzes wird in den zukünftigen eine zunehmende bedeutende Rolle einnehmen. Anforderungen an Energiebordnetze werden durch neue Funktionen zur Minimierung des Kraftstoffverbrauchs, Hybridisierung des Antriebsstrangs und Integration von Hochstromverbrauchern in einer frühen Phase definiert. Die Integration mehrere Spannungsebenen, Verschaltung, Dimensionierung von Speicher- und Erzeugerkomponenten und Migration von Verbraucher repräsentieren die wesentlichen Freiheitsgrade in der Auslegung. Durch die hohe Anzahl der Freiheitsgrade existieren verschiedene Auslegungsmöglichkeiten um den Anforderungen und gerecht zu werden. Dem steht eine wirtschaftliche-, kosten- und gewichtsoptimale Lösung gegenüber.
Dieser Beitrag stellt eine Methodik dar, um diesen Zielkonflikt aufzulösen.

1. Introduction

In future years, vehicle on-board electrical systems must overcome new challenges. These include new load systems, increased energy-/power demand, new functions to reduce fuel consumption and powertrain hybridization. Depending on the requirements for future vehicles, there are many options for designing on-board electrical systems. The characteristics of an on-board electrical system vary depending on the topology, load migration, sizing of components and functions of an electrical energy management. The focus is to ensure the voltage stability, minimize CO2 emissions and to implement new features that are visible to the customers. Based on economic efficiency, a cost- and weight-optimized solution, which will also meets the requirements, is sought. It should also be possible during the early development phase to evaluate variants that solve the trade-off between the cost of the on-board electrical system and the implementation of its requirements. Previous methods for developing an on-board power supply system only analyzed each of these requirements individually. Therefore, a new method including all aspects is presented below. The focus of this new method is currently on the dimensioning of a mild hybrid vehicle.However, it is possible to expand to other applications (e.g. Hybrid Electric Vehicle - HEV).

2. Requirements and conditions for designing on-board power supply systems

Certain conditions are defined to design an on-board power supply system. These conditions are explained below.

2.1 On-board power supply system topology

A central problem is represented by the choice of a suitable on-board power supply system topology. The topology depends on the different voltages applied (Figure 1). These levels are explained below:

- Low voltage (LV, 12 V),
- Medium voltage (MV, 48 V) and
- High voltage (HV, > 60 V).

Basically, today's vehicles always have a 12 V power supply system (LV). For simple start/stop systems or the integration of high power loads, an additional 12V power supply system (LV-LV) is necessary. If a hybridization (e.g. recuperation, shifting load point, electric driving) is to be realized, then it is necessary to use a high voltage level for full hybrids (LV-HV or LV-NV-HV) or a middle voltage level for mild hybrids (LV-MV). Full hybrids with high power loads that require more than 3kW power are achieved with a three voltage (LV-MV-HV) on-board power supply.

The correct choice of voltage levels is influenced in particular by the load systems and hybrid functions. Different voltage levels are connected to each other by a uni- or bidirectional DC-DC converter. Furthermore, the use of a Q-diode instead of a DC-DC converter for identical voltage levels (LV-LV) must be evaluated. An on-board power supply system includes the following components: energy storage, energy source and energy sink. These are explained in more detail below.

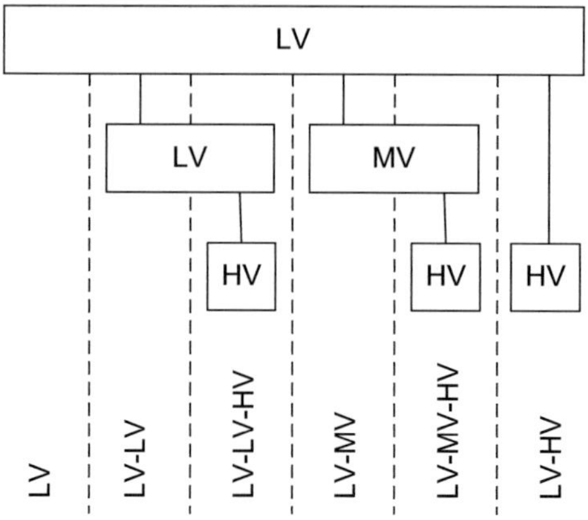

Figure 1: *Possible power supply systems in a vehicle*

2.2 Components

The requirements of an energy storage deviceare defined by the following aspects: cold cranking ability, power supply to start the engine, capacity, and power-, load- and cycling capability. Storage devices differ between their power and energy density. These are characterized by using a Ragone chart (Figure 2). Whether a storage device meets requirements depends on the technology and wiring of the cells (seriel and parallel).

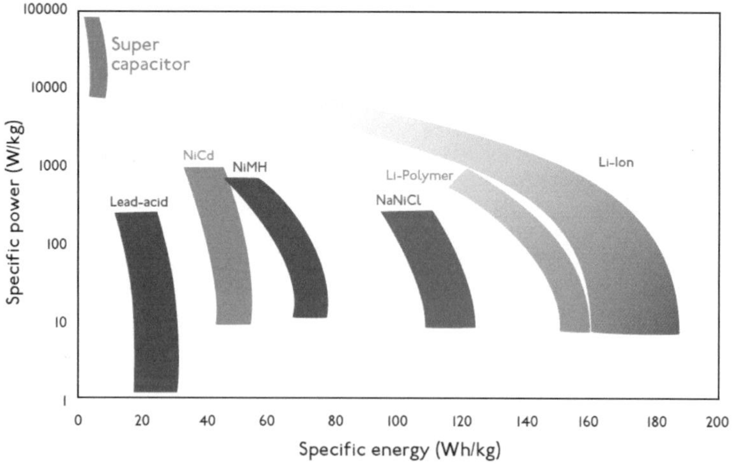

Figure 2: *Ragone-chart of different storage system technologies [5]*

In an on-board power supply system, either alternators or electric motors are used as an energy source. These are differentiated by their capability, operating mode and efficiency. A correct dimensioning of the energy source is very important in order to reach a neutral balance of energy.

At the same time, DC-DC converters are used to link different voltage levels to each other. Depending on the operating mode (buck or boost), the DC-DC converter can be used as either an energy source or sink. A DC-DC converter can be an energy source, depending on which voltage level has no alternator or electrical machine to supply the loads and the storage system. The performance of the DC-DC converter and the current dynamics have a major influence on the power and energy balance in the power supply system. For this reason, the dimensioning of these components must also be considered.

2.3 Load migration

The balance of energy and power in an on-board power supply system is influenced by the integrated loads. Loads are differentiated into the following categories:

- Powertrain Control,
- Chassis Control,
- Safety System,
- Body Control and
- Infotainment System.

If a 48 V voltage level is integrated, then there is a new degree of freedom to migrate loads between the different voltage levels. Figure 3 shows a load migration scenario for a two voltage on-board powersupply system. It is possible to optimize the power and energy balance with a suitable migration scenario.

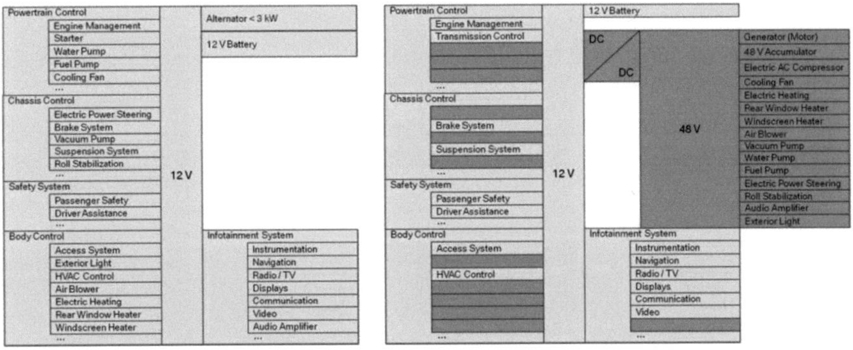

Figure 3: *Migration scenario for 12 V / 48 V on-board power supply systems [6]*

2.4 Vehicle energy management

Vehicle energy management differs between electrical energy management, powertrain management and thermal management (Figure 4). A detailed explanation of these management functions is shown in [1].

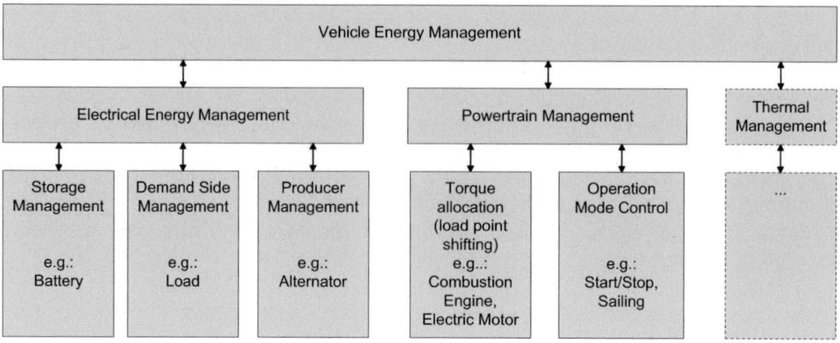

***Figure 4:** Schematic diagram of vehicle energy management [1]*

In this work, electrical energy management functions are used only to improve the voltage stability, as will be shown in section 3.3. The following options can be used to improve voltage stability with electrical energy management functions:

- reduce the power of the electrical machine in critical situations (e.g. recuperation or charging the battery),
- define a battery charge limit (reserve for recuperated energy),
- define a battery discharge limit (reserve for peak power),
- degrade or reduce the power demand of loads,
- short term reversal of operation mode of the DC-DC converter,
- and additional functions.

3. Methodology for the design of electrical energy on-board power supply systems on the example mild hybrid vehicles (12 V / 48 V – multi-voltage vehicle on-board power supply systems)

Presented in chapter 2 were the important degrees of freedom, which are responsible for a large number of possible variations of on-board power supply systems. These include, on the one hand, the design of the power system topology and the dimensions of the main components (alternator, DC converter, storage devices). On the other hand, the development of an appropriate electrical energy management (EEM) offers a variety of influences that enable an energy efficient and stable on-board power supply operation.

The first part of this chapter will present an evaluation metric for energy network topologies, specifically for mild hybrid vehicles with 12 V / 48 V multi-voltage on-board power supply systems, in order to optimize them in further steps. The second part addresses the possibility to improve voltage stability with the help of an appropriate electrical energy management. Here, an approach for the development of an adaptive EEM is briefly discussed. The combination of these two methods is the holistic approach for the design of on-board power supply systems for future vehicles.

To optimize the on-board power supply system for an energy efficient operation is of overriding importance. The control strategy for the power unit is specified and not a degree of freedom.

3.1 Simulation environment as a basis for the development of electrical energy on-board power supply systems

The evaluation of the electrical energy network topologies and the research activities to understand the effect of the EEM are based on simulations of the on-board power supply system and the powertrain. Figure 5 shows the simulation environment of a mild hybrid vehicle. The on-board power supply model consists of a starter-generator, batteries for 12 V and 48 V and a DC-DC converter to couple the voltage levels and electrical loads. When an on-board power supply model is coupled with a powertrain model, it then consists of a longitudinal dynamics model, a driver model and an integrated driving scenario.

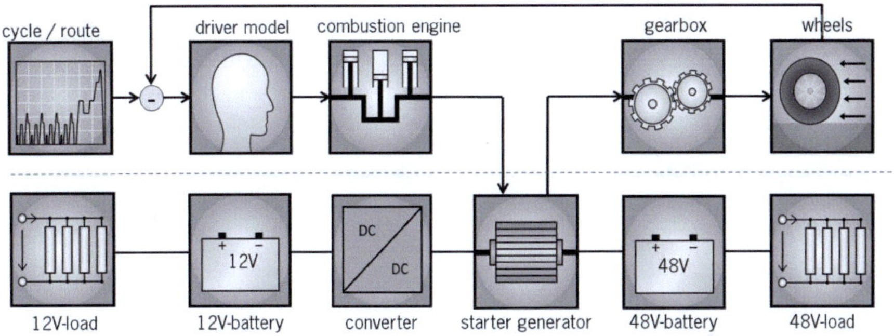

Figure 5: *Structure of the mild hybrid vehicle simulation environment*

While a self-contained modeling of the on-board power supply system is sufficient to optimize the power network topology, a driver and a powertrain simulation is needed to analyze the effect of the EEM. The integrated driving and load scenarios originate in test bench and vehicle measurements.

3.2 Evaluation concept for the design of on-board electrical energy power supply systems and for the dimensioning of its main components

The mild hybrid structure in the following figure is the basis for the described research activities.

A structured evaluation of vehicle on-board power supply systems is the basis for optimization. The use of multi-voltage on-board power supply systems offer new challenges in terms of the addressed issues. Because of the high number of degrees of freedom in the design process, it is relatively complex to develop an adequate and clear on-board power supply analysis for the use of an optimizing algorithm in further works.

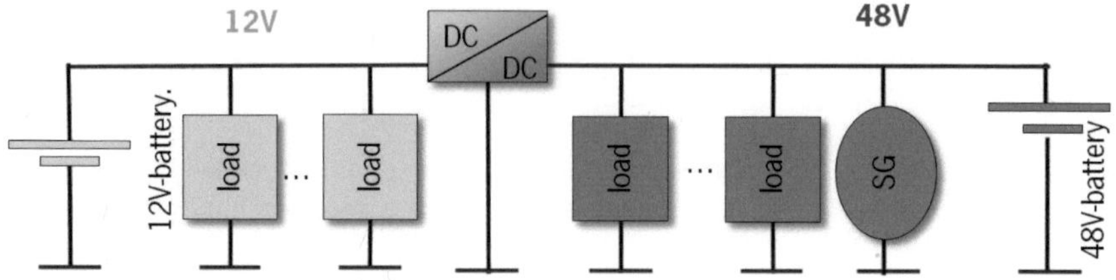

Figure 6: *Overview of the investigated mild hybrid structure*

The requirements of the on-board electrical energy power supply system are highly depend on the class of vehicle. Because of this reason, an evaluation concept had to be designed for different individual requirements. In recent months, a power supply evaluation concept was developed at the Institute of Automotive Technologies Dresden (IAD). This concept uses several established evaluation criteria. The goal was to merge these criteria together into a points system. In the end, there is a characteristic number to compare different electrical energy power supply systems. The development is focused on the optimization of a multi-voltage on-board power supply system with two coupled voltage levels (12 V and 48 V). The layout of the automotive wiring harness and other related are not part of the research at the moment. Naturally, it is possible to adapt this evaluation concept to other specifications of the on-board power supply system development.

3.2.1 The influence of different dimensioning of the on-board power supply system components

According to section 2.2, different possibilities exist for dimensioning individual components of the vehicle on-board power supply system.
An important group of components are the electrical sources. They provide electrical energy by an energy conversion. One part of this group is the alternator (possibly a starter-generator). The alternator is coupled mechanically with the crankshaft to convert kinetic energy into electrical energy. In the present consideration of the multi-voltage on-board power supply systems, the alternator is positioned at the 48 V voltage level. The second electrical source in such a system is the DC-DC converter. It interlinks both voltage levels. Here, the DC-DC converter works in boost, buck or bidirectional mode. In a way, it can also work as an electrical load. In this case, the DC-DC converter uses the energy in one part of the multi-voltage power supply system to supply the other part of the system.
For example, a buck converter is an electrical load in the 48 V level and an electrical source in the 12 V level. The storage devices are another component group. They play an important role for the voltage stability of the on-board power supply system. They supply the vehicle with electrical energy during phases of motor stop or insufficient alternator performance. In this way, there is no noticeable power loss in the electrical system and every necessary operation is working. Because of the limited

dynamics in the alternator controller, storage devices are needed for buffering transient loads.
In multi-voltage on-board power supply systems, the load allocation onto the two voltage levels is an essential influencing factor for the performance of the power supply system. It is more complex than in the conventional 12 V system. At first glance, it seems that it would be appropriate to bring all high power loads onto the 48 V voltage level for better voltage stability and a more energy efficient operation. However, the issue is not as trivial as it seems. Costs to adapt loads onto 48 V play an essential role for the allocation decision.

To change the properties of main components has an important impact on a large number of on-board power supply system characteristics. These are characteristics which have an affect on performance. Amongst other things, this includes the voltage stability, the efficiency of mild hybrid functions (recuperation, stop-start, etc.) and the battery load. The latter has a significant impact on the State of Health (SoH) of the batteries and thus influences the change in performance over the operation time. Furthermore, the dimensioning of the components has a significant effect on the weight of the on-board power supply system, the required installation space and also on the costs for production and development.

It is possible to focus on every discussed point for an optimization of the on-board power supply system, but attention should be paid to the different trade-offs.

For example, the trade-off between a cost and a performance optimized on-board power supply system. The optimization in terms of highest possible overall performance, for representing various on-board power supply features and high voltage stability, is not conducive to keeping costs within reasonable limits. Because of the goal to develop a system with high performance, a powerful alternator, a powerful DC-DC converter and a corresponding battery is needed.
For this reason, it is necessary to establish the evaluation of the on-board power supply system with a focus on the user requirements or the demands of a particular class of vehicle. This is done with the help of weighting evaluation criteria.
Another instrument, especially for an energy efficient vehicle, is the operations strategy or the drive management. Naturally, this influences the behavior of the electrical on-board power supply system. Primarily, in the field of the discussed mild hybrid vehicles with a 48 V voltage level, the on-board power supply system is under higher pressure than in older 12 V power supply systems, where the supply of electrical loads was the main focus. In stop-start phases, the whole electrical supply must be managed by the battery system. While in times of recuperation, the batteries need to have enough capacity to absorb the gained energy. This clearly shows that the requirements for future batteries are higher than in the past years.
For the evaluation concept that is present on the next chapters, the operation strategy plays a minor role. It is taken as given and not as a part of the actual research activities in this field. Thereby, simple control strategies for stop-start and recuperation are deposited in the drive cycles. With the help of the EEM, influence on the operation strategy is possible. It is shown in section 2.4.

3.2.2 Evaluation approach and criteria

Table 1 shows an overview of the evaluation criteria used. A distinction is made between two approaches. On the one hand there are so-called analytical criteria and on the other hand, the qualitative criteria.

Analytical evaluation criteria

Analytical criteria are characterized by factors which enable a comparison of different on-board power supply systems to each other. These criteria include, among others, the charge condition or the average recuperation current. The latter gives information about the maximum recuperation potential during a defined drive cycle. There are also disadvantages of these criteria. They are highly dependent on the used loading and driving cycles.

Qualitative evaluation criteria

The qualitative criteria are based on the assessments of the corresponding departments. Given that the evaluation is done in an early development phase, many details are not yet explicitly defined. So the evaluation of a criterion is really only an estimate. Thereby, the previous experiences from competent employees are very important in order to estimate as closely as possible to reality. The evaluation is always performed in comparison with a reference on-board power supply system. This can be, for example, a conventional 12 V on-board power supply system or any other multi-voltage power supply system. In the presented approach, the evaluation follows a point system. More information will be given in the next chapter.

Table 1: *Overview of evaluation criteria used*

Analytical criteria	Qualitative criteria
Charge condition	Change of weight
Degree of efficiency	Cost development
Power supply without consumption influence	Complexity
Average recuperation current	Constructed space
	Costumer experience (potential for new automotive functions)
	Battery requirements

3.2.3 Evaluation concept

The goal of this concept is the consolidation of the two evaluation criteria which are described in the last chapter. At the end of the evaluation, there should be only one classification number for evaluating the investigated on-board power supply systems.

For this, a point system is introduced. Figure 7 shows an overview of the developed evaluation concept.

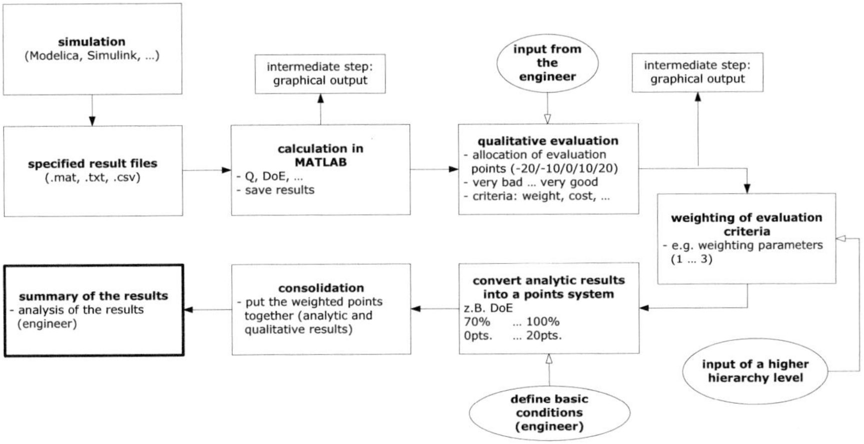

Figure 7: *Conceptual figure of the evaluation methodology*

In the first step, the researched variants of the electrical on-board power supply systems will be simulated. The simulation tool is not specified. It is only important to use a defined result file. At the IAD, the chosen programming language is the object-orientated language, *Modelica*. The simulation is performed using *Dymola* and the *IADLib* (the institute's own library for on-board power supply systems) [7]. From these results, the analytical criteria can be calculated in the next step. Afterwards, the results need to be analyzed a first time. If the results don't meet the requirements, a new simulation is necessary. The third step is the qualitative evaluation of different variants. In step four, the evaluation results can be weighted and, therefore, adapted to one's own preferences. Figure 8 shows a transformation of the analytical results into a points system. This is necessary to consolidate both evaluation approaches (analytical and qualitative).
With the help of the transformation into a points system, the analytical results are presented in a way that they can be subsequently combined with the qualitative results, which are also given in a points system. This makes it possible to present a compacted overall result. The conversion into the points system is as follows:
Before the conversion starts, ranges of validity are defined. Possible points are scattered about this range. Poor results receive less or no points as opposed to very good results. The transformation follows a comparison of the analytical results with an interpolated range of validity.
The final result offers a good possibility to compare different variants of on-board power supply systems with more than one evaluation criteria.

In [2] the evaluation concept is described at full length.

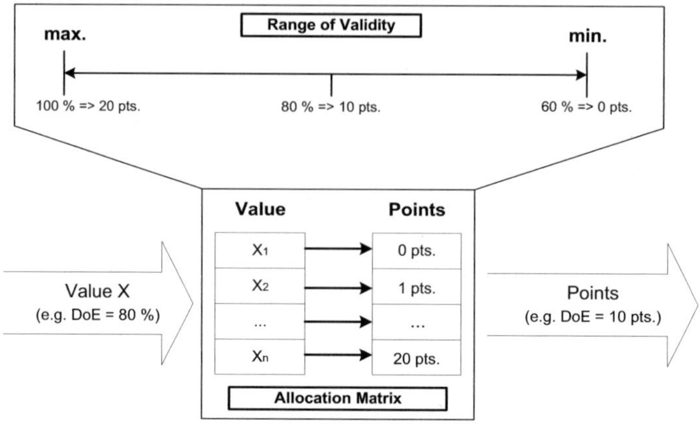

Figure 8: *Transformation of the analytical results into a points system (exemplary range of validity)*

3.3 Optimization of the voltage stability of qualified electrical on-board power supply systems by electrical energy management

The previous chapter was an introduction into the evaluation concept for energy efficient multi-voltage on-board power supply systems. The improvement of the voltage stability through the influence of the EEM will now be described. An energy and cost optimized electrical on-board power supply system cannot guarantee adequate voltage stability in all situations. As presented in Figure 10, it is possible to shift the voltage stability into an area that meets the requirements of the on-board power supply system. After the optimization with focus on energy efficiency, the next step is to find a possible EEM configuration to improve the voltage stability of the optimized and chosen on-board power supply systems.

Voltage interruptions or voltage increases in on-board power supply systems are critical when they go out of the permitted range. Within the lower and upper voltage limits, an unrestricted operation is possible. If the voltage moves out of this range, measures must be taken to avoid defects or damage in the on-board power supply system. If possible this situation should be avoided before it occurs. Here, in addition to the design of the electricalnetwork topology, the EEM plays an important role.

Critical voltage variations are possible at both voltage levels (12 V and 48 V) in multi-voltage systems. For voltage stability, the higher voltage level is the focus of research because of the migration of main components and loads onto this level. One main component is the alternator or the starter-generator. High power loads are typical for a 48 V migration, such as active roll stabilization. Another reason for higher and critical voltage variations at the 48 V voltage level is the influence of mild hybrid functions on the on-board power supply system. Functions like stop-start, stop-on-a-move and brake energy recovery, that set a higher target voltage at the alternator, are critical for voltage stability.

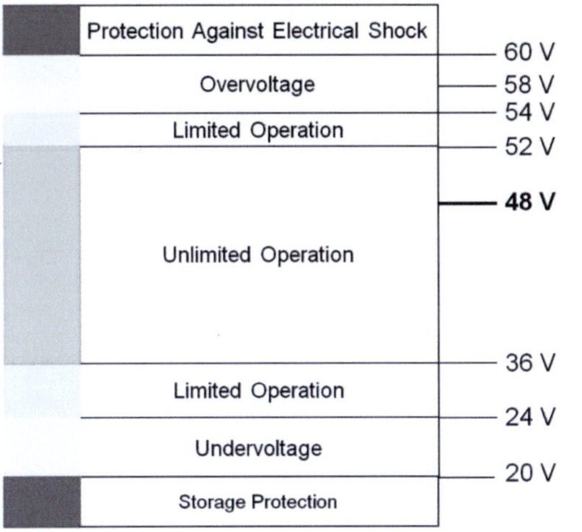

Figure 9: *Definition of the static voltage range in 48 V on-board power supply systems [6]*

Figure 9 shows the static voltage range in 48 V on-board power supply systems. Based on these limits, the voltage stability of a 48 V or a multi-voltage level on-board power supply system need to be investigated. Metrics for the evaluation of 12 V and 48 V voltage levels are presented in [2] and [3]. The main difference for the evaluation is the calculation of the stability reserve R_{St} and the degree of stability G_{St}. In contrast to the evaluation in 12 V power supply systems, the evaluation in 48 V systems uses no nominal voltage U_N, but rather a voltage range which is limited by U_{min} and U_{max}.
The described methods to evaluate the voltage stability of an on-board power supply system are also relevant during the EEM development. In section 2.5, possibilities to influence the voltage stability with the EEM are discussed. One of these functions will be briefly presented in the following. [1] presents an adaptive electrical energy management for 48 V on-board power supply systems.

This method especially helps control the load and charge characteristics of the battery. Enough buffer capacity must be available because it is not possible to know how much energy the loads need and how much power the alternator is able to produce at any time. The driver desired torque is the most significant disruptive factor. Through a predictive diagnosis of the load management, the needed load power can be determined.
The battery management measures voltage, current and temperature. The internal resistance and the open-circuit voltage can be derived from these results. From these parameters, necessary battery conditions, such as State of Charge (SoC), State of Health (SoH) and State of Function (SoF), can be determined. With a model based power prediction, the permitted charge and discharge power is set. With the help of the reserved load power and the battery performance, the power consumption and output of the starter-generator and the 48 V load are reduced to ensure compliance with the current and voltage limits.

Furthermore, the EEM allows for the possibility to migrate new loads into an existing on-board power supply system in order to integrate new functions. Thus, no modifications are needed and additional development costs can be saved.

4. Results

Figure 10 illustrates the potential of a combination between the design of an electrical power supply system and an adequate electrical energy management. After the selection of an energy-optimized on-board power supply system with the use of the multi-level evaluation method (see also [2]) is found under consideration of the given requirements, it is important that this system is checked in regard to the voltage stability. [3] describes a method to evaluate the voltage stability in conventional power supply systems with two characterization factors (degree of stability G_{St} and stability reserve R_{St}). An adaption of this method to 48 V power supply systems is presented in [1].

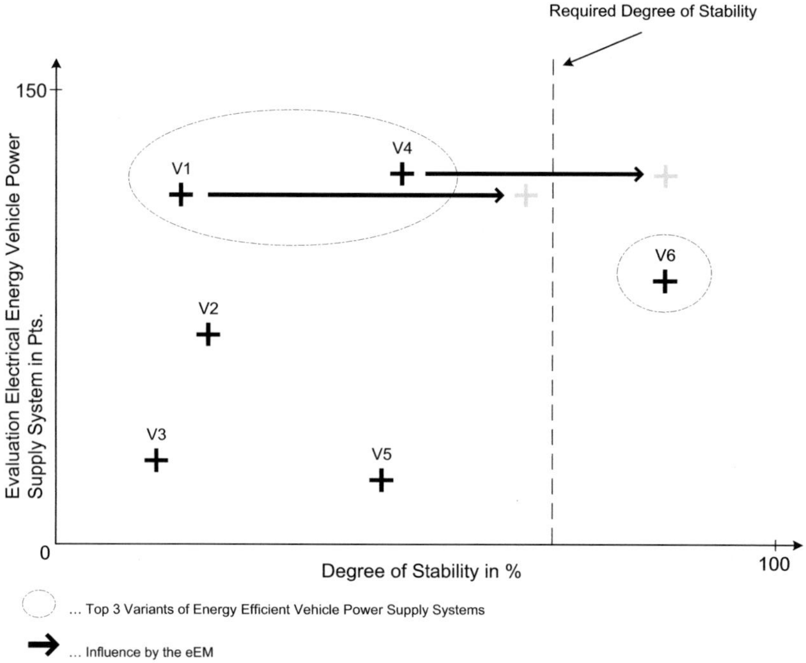

Figure 10: *Influence of the electrical energy management in regard to improving the voltage stability of energy-optimized electrical on-board power supply systems*

After the selection and screening of adequate on-board power supply systems, they are plotted in Figure 10. The y-axis describes the evaluation of the on-board power supply system with focus on energy efficient operation, which is plotted in a points system. In section 3.2 you can find a more detailed explanation. On the x-axis, the voltage stability is plotted with regard to the degree of stability. The EEM design enables an adjustment on the x-axis into an area of higher voltage stability. There are many possibilities to improve the voltage stability with the help of the EEM. The possibilities are described in the chapter before. They can be divided into three instances: battery management, electrical load management and electrical source management.

Crosses mark the evaluation results from 6 different electrical on-board power supply systems in figure 10. With regard to energy efficiency, variants V1, V4 and V6 are the highest rated. The dashed line parallel to y-axis marks an area of voltage stability which must be reached in order to meet the minimum requirements of this example on-board power supply system development. V6 reaches the requirements without additional EEM measurements. With the help of adequate EEM functions, V4 can be shifted into the required voltage stability area. Therefore, this is the best variant to meet the requirements. Also, with an optimized EEM, V1 cannot be shifted into the stable area.

5. Summary

In this thesis, a holistic approach was presented for the design of on-board power supply systems in an early development phase for future vehicles. The main problem lies in the definition of an adequate electrical on-board power supply system in regard to the user requirements. With the help of analytical and qualitative criteria, the presented metric can be used to limit an optimal solution space within the design degrees of freedom. The main focus is on the consideration of economic efficiency, energy efficiency and voltage stability.

Literature

[1] V. Schäfer, Adaptive Energy Management for 48V Power Networks, Elektrik/Elektronik in Hybrid- und Elektrofahrzeugen und elektrisches Energiemanagement, Bamberg, 2014

[2] S. Löwenberg, Model-based multi-level evaluation methodology for conceptual design of 48V/12V Multi Voltage Vehicle Power Supply Systems, Elektrik/Elektronik in Hybrid- und Elektrofahrzeugen und elektrisches Energiemanagement, Bamberg, 2014

[3] J. Klötzl, Stability in Automotive Pwer Nets: Definitions, Algorithms and Experimental Validation, Power Electronics and Applications, Birmingham, 2011

[4] S. Kehl, A. Radon, A. Klinkig, Elektrische und elektronische Komponenten im Fahrzeug – 48 V Bordnetz – Anforderung und Prüfung, 2013

[5] P. Van den Bossche, The Cell versus the System: Standardization challenges for electricity storage devices, EVS24 International Battery, Hybrid and Fuell Cell Electric Vehicle Symposium, Stavanger, 2009

[6] T. Dörsam, D. Grohmann, S. Kehl, A. Klinkig, A. Mai, O. Sirch, A. Radon, J. Winkler, Die neue Spannungsebene 48 Volt, Baden Baden, 2011

[7] T. Schwan, L. Morawietz, B. Bäker, Entwurf einer Li-Ion-Fahrzeugbatterie auf Basis einer Fahrzeug-Bordnetzsimulation, Dresden, 2010

15 Two voltage battery as one storage solution for a 12 V/48 V power supply

Dipl.-Ing. **Ronny Petersohn**, Prof. Dr.-Ing. **Bernard Bäker**, Technische Universität Dresden, Dresden, Dipl.-Ing. **Rainer Falsett**, Daimler AG, Böblingen, Germany

Abstract

The electrical requirements of power and energy in vehicles continue to increase as a consequence of constant development in driving safety, comfort, and the electrification of automotive components. Finally, the vehicle power supply has reached its physical limits. To satisfy these risen demands to on-board power supplies, several German automobile manufacturers decided to develop a second voltage level. Thus, the 12-volt level is expanded with a 48-volt level [1]. As a result, an additional accumulator is used. Considering that these implicate the disadvantage of increased installation space, weight, and cost; an electrical storage on lithium-ion basis in the form of a battery with a medium voltage tap is to be used. However, a disadvantage is the unbalanced load of the storage, which makes a charge equalization circuit necessary. In cooperation with the Daimler AG the usability of such a dual-voltage battery is examined within the public funded research project EFA 2014/2 (Energy Efficient Drive 2014/Part II). For this purpose a model was created which is used for interpretation of such a two-voltage-battery with charge balancing. In this paper the model as well as a test bench, which is used to validate the simulation results, are presented.

Kurzfassung

Die Belastung des Fahrzeugenergiebordnetzes steigt seit Jahren kontinuierlich an. Weiterentwicklungen in Richtung CO_2-Ausstoßreduktion, Fahrsicherheit, –komfort und der Elektrifizierung von Fahrzeugkomponenten haben daran einen wachsenden Anteil. Dabei kommt das Fahrzeugenergiebordnetz immer näher an seine physikalischen Grenzen. Um diesen gestiegenen Anforderungen an das Bordnetz gerecht zu werden, wurde die Entwicklung einer zweiten Spannungsebene von einigen deutschen Automobilherstellern beschlossen. Somit wird die 12-Volt Ebene mit einer 48-Volt Ebene erweitert [1]. Dies hat zur Folge, dass zwei elektrische Speicher Verwendung finden. Da diese die Nachteile von erhöhtem Bauraumbedarf, Masse und Kosten mit sich bringen, soll ein Speicher mit Mittelspannungsabgriff hierfür eingesetzt werden. Großer Nachteil eines solchen Speichers ist jedoch die unsymmetrische Belastung, welche eine Ladungsausgleichsschaltung notwendig macht. In Zusammenarbeit mit der Daimler AG wurde die Realisierbarkeit einer solchen Zweispannungsbatterie innerhalb des öffentlich geförderten Forschungsprojekts EFA 2014/2 (Energieeffizientes Fahren 2014/Phase II, Erhöhung der Reichweite von Elektrofahrzeugen) untersucht. Hierzu wurde ein Modell erstellt, welches zur Auslegung einer solchen Zweispannungsbatterie mit Ladungsausgleich genutzt wird. In diesem Beitrag soll das Modell sowie ein Prüfstand, welcher zur Validierung der Simulationsergebnisse dient, vorgestellt werden.

1. Introduction of a two voltage battery

The considered dual-voltage vehicle electrical system consists of a 48-volt alternator, the associated loads and a 48-volt storage. This 48-volt network is extended with a 12-volt network including the 12-volt storage and the according loads. The interconnection of the networks is realized using a DCDC-converter. By use of such a dual-voltage vehicle electrical system, high-power load can migrate from the 12-volt into the 48-volt electrical system. As a result, an increase of the efficiency and a reduction of the cable cross sections can be realized. In addition, higher recuperation performance, an extension of the start-stop functions, as well as an electric motor operation with load point displacement, can be realized. This contrasts with the additional space required, the increased cost, and the extra weight through the use of a second storage. A two-voltage-battery can therefore contribute to a limitation of these parameters by reducing the number of the necessary storage units to only one (Fig. 1). This battery is assembled with a set of 14 in series connected 2.3 Ah iron phosphate round cells. However, for possible use in the series production vehicle, this capacity is insufficient. Therefore, 20 Ah cells of the same chemistry are used. To obtain meaningful results the load currents are limited — according to the ratio of the cell capacity — to a tenth of the vehicle electrical system loads occurring. The 12-volt voltage is tapped after the fourth cell. The occurring problem is the asymmetric loading of such a two-voltage-battery.

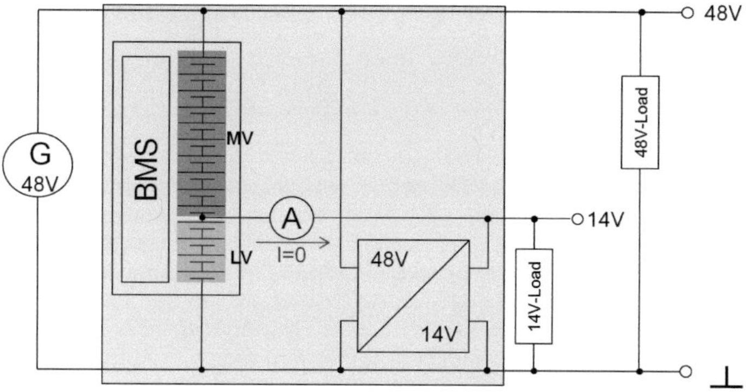

Figure 1: *Two-voltage-battery*

The 12-volt module (Fig. 1, LV) must also cover the 48-volt loads in addition to the 12-volt loads, so it's deeply discharged. This can lead to a stronger aging of the module in relation to the MV module (Fig.1, MV) and the start ability of the vehicle cannot be guaranteed, since the 12-volt module cannot be fully charged. A tangible solution is to use the built-in DC-DC converter of the dual-voltage vehicle electrical system, which prevents the discharge of the storage (zero current control) and can balance the charge if necessary. Ideally, the battery behaves like a symmetrically charged battery. Therefore, in the following section a simulation model is presented, which serves as a tool for designing a lithium-ion two-voltage battery with charge balancing. Further to this, a test bench that was used to validate the model is described.

2. Model Formulation

The simulation of the dual-voltage vehicle electrical system is done using the object-oriented modeling language Modelica, respectively the graphical development environment Dymola, and the vehicle electrical system simulation library of the Dresden Institute of Automobile Engineering. To design the compensation circuit, several requirements must be met regarding the model. The state of charge of the cells must be able to be simulated with given loading scenarios and cell states. It should also be possible to enter a variety of specifications for the design, e.g. the time at which the cell must be discharged, the minimum state of charge, and the time after which a balancing process must be completed in compliance with critical states. In addition, the operating range of the accumulator must be reliably maintained. Thus, a battery management system (BMS) has to be implemented in the model. Due to the object-oriented approach the electrical system can be implemented quickly. The parameterization of the necessary individual cells and the DC-DC converter is performed on the institute's own test benches. To model the dual-voltage battery, multiplexers are used (Fig. 2, position 1). This means that only one cell has to be simulated per module.

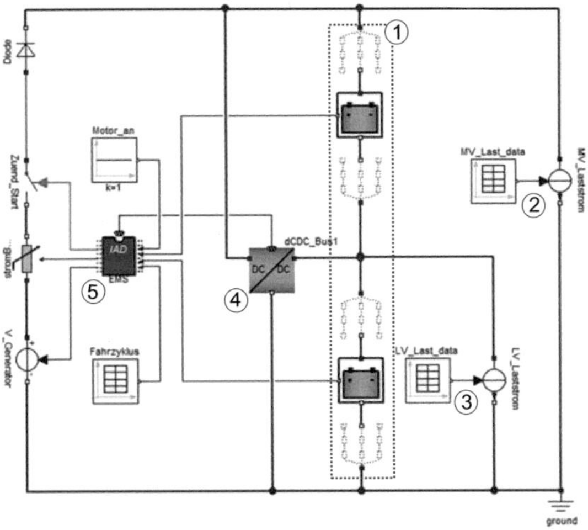

Figure 2: *Simplified simulation model*

The input parameters of a multiplexer are the number of cells in series and the cells in parallel. This allows a memory topology change only by changing these parameters and thus the model structure is preserved. However, module-internal charge differences cannot be considered with this kind of model. As a matter of priority, the asymmetry between modules is more important so the interpretation of module-internal information may be waived.

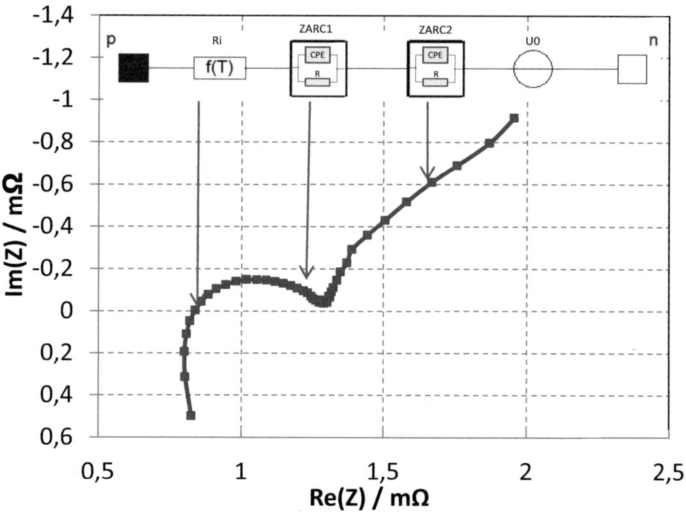

Figure 3: *Impedance spectrum and electrical circuit model*

The individual cells themselves are represented by an impedance based simulation model, which can be seen in figure 3. The parameterization of such a model is done by the electrochemical impedance spectroscopy. In this case, the impedance spectrum is fitted with the aid of the equivalent circuit, whereby the current and voltage profile of the cell can be modeled. In the model itself the established parameters in dependence on temperature and state of charge are deposited. These can be replaced without much effort, so an easy change of the storage technology is possible. Also the capacity, the state of charge (SOC), and the state of health (SOH) can be changed. For the investigation of different load scenarios, controlled current sources are used, — where each voltage level has its own (Fig. 2, position 2 and 3). The load profiles are deposited as mat-files.

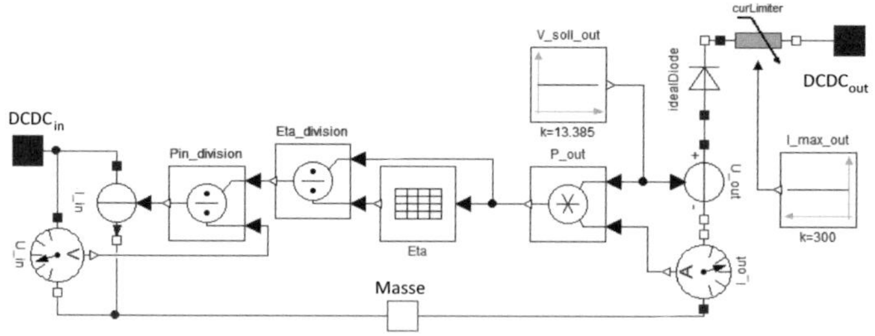

Figure 4: *Buck converter model*

The modeling of the DC-DC converter is realized by an energetic approach (Fig. 4). This relationship is shown in Equation (1). At the output of the converter operates a voltage source, which is adjusted to the desired voltage.

$$P_{in} = \frac{P_{out}}{\eta} \tag{1}$$

$$I_{in} = \frac{V_{out} \cdot I_{out}}{\eta \cdot V_{in}} \tag{2}$$

An ammeter is integrated in the same branch. From those two blocks, the output power is determined by a simple multiplication. The result is set into relation with the deposited efficiency factor leading to the required input power. In the next step this power is divided by the current input voltage. This leads to the input current for the power source (Equation (2)). In addition to the change in the output voltage, it is also possible to vary the output current of the converter within the physical limits. To control the load balancing process, in addition to maintaining the operating limits, a state machine was created with *MathWoks Stateflow*. For this purpose it was necessary to implement the *Dymola* model as *S-Function* in *Matlab / Simulink*. Using a graphical user interface (GUI), the user is still able to change all of the input parameters of the model. In principle, two operating strategies are possible: the previously introduced zero current control and a converter optimal control. Additionally, considering the possibility of using current and voltage regulated converters, 4 operating strategies are obtained. The current-controlled converter is adjusted to the power required by the 12-volt voltage level. When using the voltage-controlled converter, however, an attempt is made to settle the output voltage in the same proportion like the voltage of the alternator, so that the performance requirements from the 12-volt vehicle electrical system are covered by the DC-DC converter. All operating strategies may be realized and examined using the model. For further information, it is referenced to [2].

3. Test Bench

The model allows the study of different charge balancing options for different battery topologies. To assess the validity of the simulation results, real measurements are required. For this purpose a test bench has been realized, which allows a wide range of tests comparable to the model. For example it should be possible to change the topology of the battery and to test different module-internal and module-external load balancing options. In addition, the test bench should react autonomously when limits are violated, so that the presence of staff is not required continuously during the test periods. For this reason, we choose a compact and reconfigurable control and monitoring system, a compactRIO (cRIO), from National Instruments, which can operate as a stand-alone system without a host computer. The tasks of the cRIO can be divided into two parts. Firstly, the cRIO acts as a BMS and thus it is responsible for the limit value monitoring and the charge balance. Secondly, it serves to control the electric source, which is intended to reflect the 48-volt alternator, and the two electric loads, which represent the respective loads of the vehicle electrical system. Figure 5 illustrates this relationship. There are different modules with which the cRIO has been expanded to take over the various tasks. Currently 6 modules have been installed. The system can be expanded up to 8 modules. The NI 9211 and NI 9213 are used for temperature control, each cell is individually monitored. The cell voltage is monitored by module 9220. Each channel of this module must remain within ±10.4-volt to ground. An operational amplifier had to be used to monitor the whole battery whose maximum voltage is 50.4-volt. The current is measured with current transformers. Its values, as well as the measured values of the source and loads, are read by the module NI9205. The control of the test bench components and the output of the measured values to the built-in breadboard panel meter are realized by using the NI 9264 module. For the safe switching of the power circuit and controlling the balancing circuit the NI 9403 module is used.

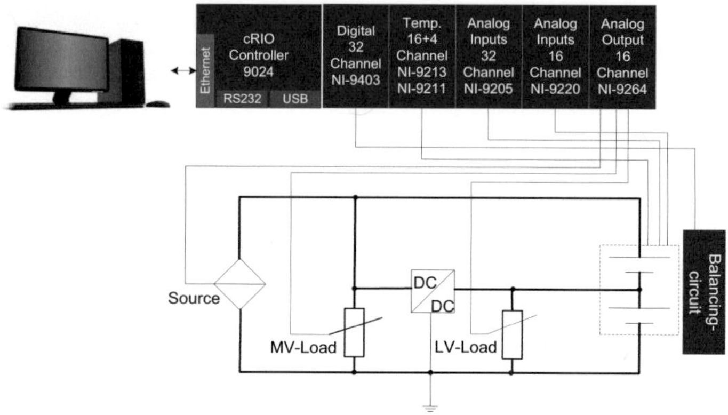

Figure 5: *Structure test bench*

To change the topology of the battery as well the load balancing options, the battery is realized as a breadboard (Fig. 6). This makes it possible, for example, to expand the capacity of the 12-volt module by adding more parallel strands. Up to 5 strands can be installed.

Figure 6: *Breadboard of the two-voltage-battery*

In addition to the extensibility of the battery, key parameters can be read directly on the breadboard. Errors can also be illustrated with the help of the status indicator, which is also attached to the breadboard.

The system is completed by a 19" rack, which includes the source, and loads additionally to the external module balancing, as well as the contactors, which can be switched via the *cRIO*. The programming of the cRIO occurs via *LabVIEW RT*. For this, the development environment of LabVIEW RT and LabVIEW FPGA is required.
There are two modes to choose for the program development on an RT (real time) target system with FPGA, which is represented by the *cRIO*. The first is the realization with the Scan Interface. The second is another programming via the FPGA interface. While a program in the scan interface exclusively uses the processor of the cRIO for execution of the code, the programming using the FPGA interface also uses the integrated FPGA. As a result, processes can run parallel. Therefore, the second method was used to control the modules and to realize the data recording, as well as the monitoring of the parameters. The data storage and transmission of the measured data have been realized on the real time controller. Also a user interface has been created for the operation of the test bench, which can be seen in figure 7. The black-framed box 1 identifies the area which is provided for user input. In addition to the name of the measurement, the driving cycle, the number and connection of the cells, as well as their start-SOC, can be specified. The appearance of the user interface is modeled according to the structure of the breadboard. The electrical source is located at position 2, which represents the alternator. Below that, position 3 shows the instantaneous profile of the current and voltage of the source. Also the characteristics of the DC-DC converter are displayed in position 4. The LV and MV module is visualized as loading bars under 5. The individual voltages and the maximum temperature of the module can be found to the left. In the right part of the user interface the LV and MV load are shown. The diagram 6, in between, shows the currents and voltages curves. The SOC curve is located at position 7 in the upper right corner.

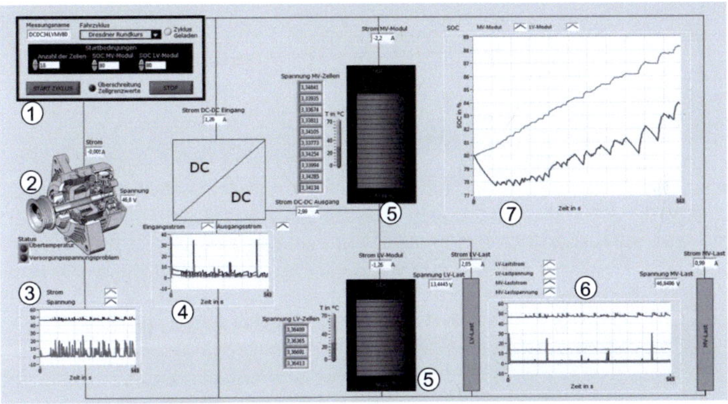

Figure 7: *User interface of test bench control*

4. Results

To validate the accuracy of the model, a test cycle is defined in addition to the use of classical driving cycles such as the New European Driving Cycle (NEDC), as can be seen in figure 8. An important aspect of this cycle is the study of the dynamic behavior of the storage. As described in section 2, the cells are assumed identical in the model. Although the choice of cells for the battery takes the minimization of deviations into account, there may be variations in the ohmic resistance, in particular when fixing the individual cells. Furthermore, charge state differences are expected due to the very flat open circuit voltage characteristic of the used $LiFeP0_4$ cells.

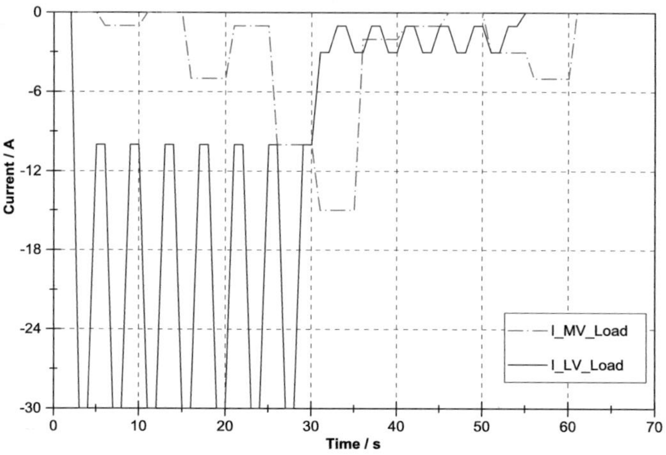

Figure 8: *Test cycle*

The results, which are shown in figure 9, show the voltage curve during the test cycle in comparison to the simulated curve without the use of the DCDC converter. The simulation and measurement agree well. Differences are particularly apparent from the discharging because of the current direction dependence of the internal resistance. From the 30th second the voltage changes are identical, only a small offset can be seen. This is due to voltage drops in the electronic source, which do not occur in the model.

The functionality of the zero current regulation was also shown using the test cycle. For this purpose, the output voltage of the DCDC converter is adjusted to the appropriate SOC desired open circuit voltage.

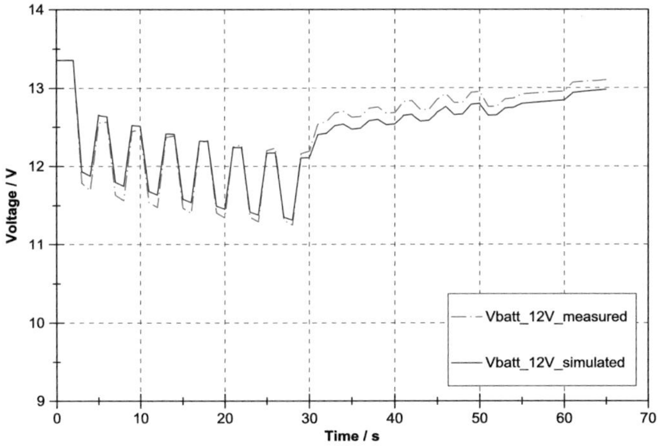

Figure 9: *Voltage curve of the 12-Volt module*

Figure 10 shows the current curves of the used 48/12-volt DCDC converter and the 12-volt and 48-volt modules. It can be seen that the 12-volt load is covered by the DCDC converter. In the present case, the DCDC converter provides the required power, so that the current coming from the 12-volt module is approximately zero. At powers which cannot be delivered by the converter, the energy consumed by the LV module has to be recharged at a later time, when the power demand is lower.

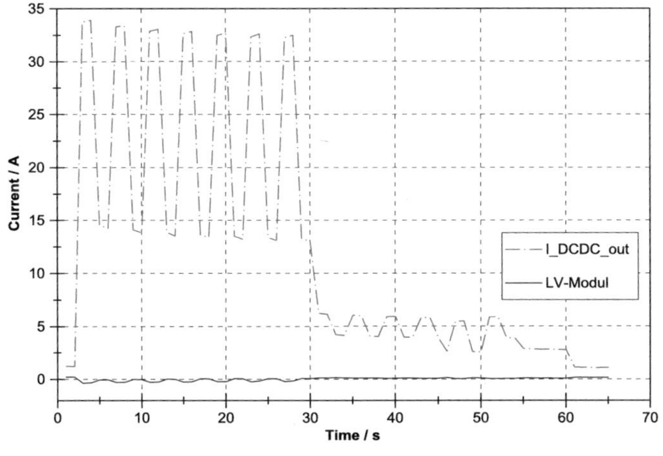

Figure 10: *Current profile of the 12V module*

5. Conclusion

In this paper, a model was presented for designing a two-voltage battery. Significant problems were the asymmetrical load of the storage and the validation of the model by using a new test bench.
Using the model and the test bench, it was possible to understand the functionality of a two-voltage battery and investigate suitable compensation options and operating strategies. The Main focus in the design of a charge compensation and the associated operating strategy are its control effort, cost, voltage stability and the Ah throughput of the single cell-module. Through the validation of the model the accuracy of the model was shown, as well as the feasibility of the two-voltage battery.

Acknowledgment

The authors wish to thank the Daimler AG whose support of this project has made it possible in the first place. Furthermore, we want to thank the students who have contributed to the realization of the project.

References

[1] Dörsam, T.; Sirch, O.: Die neue Spannungsebene 48 Volt im Kraftfahrzeug. ATZelektronik, Feb. 2012.

[2] Urs Böhme, Christian Buhlheller, et. al., Innovative Vehicle Power Supplies and System Components. eehe, May 2014 in Bamberg.

16 Hardware Test Bench for the Evaluation of E/E-Architectures with Respect to Partitioning of Vehicle Functions and Energy Efficiency

Dipl.-Ing. **Gregor Walla**, M.Sc. **Andreas Barthels**, B.Sc. **Dirk Gabriel**, Dipl.-Ing. **Florian Ruf**, M.Sc. **Michael Winter**, Prof. Dr. **Andreas Herkersdorf**, Technische Universität München, Munich, Dipl.-Phys. **Hans-Ulrich Michel**, BMW Forschung und Technik GmbH, Munich, Germany

Abstract

The e/e (electric/electronic) architecture of a modern vehicle consists of a complex distributed network of electronic control units (ECUs) interconnected by various communication buses. The electrical power consumption of the overall systems is not neglectable anymore since it negatively affects the fuel consumption and CO2 emission of a vehicle. The electrification is further increasing with every vehicle generation and energy efficiency is becoming one of the main design goals. In this paper a prototypical hardware test bench is presented that can be used for the evaluation of e/e architectures with respect to energy efficiency. Different power management techniques can be implemented and compared with each other. Experimental results indicate the potential power consumption savings for different test scenarios and the effects of the function partitioning on the overall power consumption.

Kurzfassung

Die E/E (Elektrik/Elektronik) Architektur heutiger Fahrzeuge besteht aus einem komplexen, verteilten System aus verschiedenen Steuergeräten, die durch zahlreiche Kommunikationsbusse miteinander verbunden sind. Die elektrische Leistungsaufnahme eines solchen Systems ist nicht mehr zu vernachlässigen, da sie sich negativ auf den Kraftstoffverbrauch und die CO2 Emission eines Fahrzeugs auswirkt. Mit jeder Fahrzeuggeneration nimmt die Elektrifizierung immer weiter zu. Energieeffizienz ist mittlerweile eines der wichtigsten Entwicklungsziele. In diesem Beitrag wird ein Prüfstand vorgestellt, der zur energetischen Evaluierung von E/E Architekturen verwendet werden kann. Es können verschiedene Energiesparmechanismen implementiert und miteinander verglichen werden. Unsere Versuchsergebnisse zeigen die Energieeinsparpotenziale für verschiedene Testszenarien und den Einfluss der Funktionspartitionierung auf den Gesamtverbrauch.

1. Introduction

With customer demand for more comfort and safety features the number of electric and electronic components in modern vehicles is constantly increasing. Currently the e/e (electric/electronic) architecture of a luxury car consists of a complex distributed system of up to 70 heterogeneous electronic control units (ECUs) interconnected by several communication networks (CAN, FlexRay, LIN, etc.) [9]. The energy consumption of those components is not neglectable anymore. In order to tackle this problem, different power management concepts for automotive systems were introduced recently [2, 6, 7]. The goal is to temporarily turn off vehicle functions, ECUs, or even whole bus systems. However, the effectiveness of this power management techniques is strongly dependent on the deployment of the vehicle functions.

In this paper a hardware test bench for the evaluation of e/e architectures is presented. It consists of 17 interconnected, so-called Dummy-ECUs that represent a subset of an automotive IT architecture. The self-developed ECUs are based on an ARM CPU and support different power saving methods. The test bench is used to investigate the impact of the function partitioning on the power consumption of the system being exposed to specific driving cycles.

2. Partitioning Problem

The different functions and features of a vehicle can be modeled as functional chains. A functional chain is a directed acyclic task graph (DAG) and represents a sequence of successive calculation steps, e.g. getting sensor values or data processing. The nodes of the task graph indicate the tasks to be executed and the arcs show the communication dependencies between the tasks. An exemplary functional network is depicted in figure 5.

During the e/e development process the engineers need to decide which task will be executed on which hardware component. That is, which function/feature will be integrated on which ECU. Some tasks are bound to specific hardware components because of fixed constraints (e.g. hardware resources, location of sensors and actuators, safety and security requirements, etc.) others can be freely distributed all over the system as long as certain requirements are met (e.g. timing conditions).

With the increasing complexity of the overall system this deployment problem is not trivial and is nowadays mostly based on expert knowledge from previous vehicle generations. However, with energy-efficiency considerations in mind, the partitioning problem is getting even harder. The partitioning of vehicle functions may have a direct impact on the effectiveness of certain power management techniques.

Not all functions and features are needed during all driving situations and can therefore be switched off when they are not needed. An example might be the park distance control. This function is only needed during the parking maneuver and could always be switched off whenever the vehicle is moving faster than a certain speed. Thus, if only this function is integrated on an ECU, the whole ECU could be turned off. However, when there were also other functions partitioned to that ECU (that are required during all driving situations) this can't be done anymore.

3. Hardware Test Bench

The hardware test bench is based on a real car body and represents a subset of an automotive e/e architecture consisting of 17 interconnected ECUs. The so-called Dummy-ECUs are self-developed platforms based on a mobile ARM processor (OMAP3503). The processor is integrated on the ECUs by a commercially available GUMSTIX board [3]. The ECU with the GUMSTIX board can be seen in figure 1.

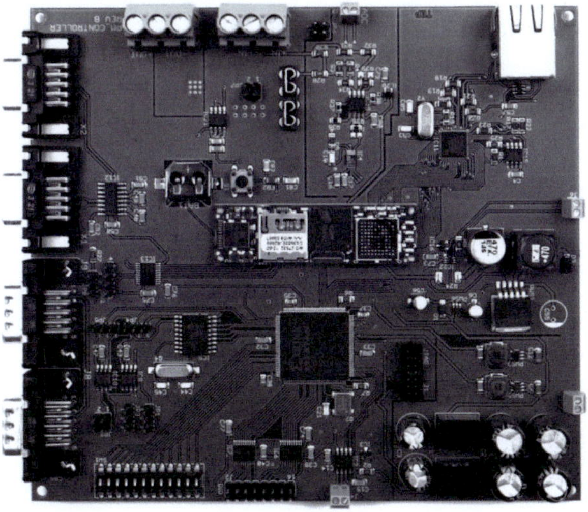

Figure 1: *Dummy-ECU with GUMSTIX Board.*

In contrast to current automotive ECUs, the ARM processors supports various power management features (e.g. dynamic voltage and frequency scaling (DVFS) or dynamic power switching (DPS) [8]) and is therefore applicable in this context. The impact of DVFS on the power consumption can be seen in figure 2. Here we measured the power consumption of the GUMSTIX board depending on the CPU utilization for three different operating frequencies.

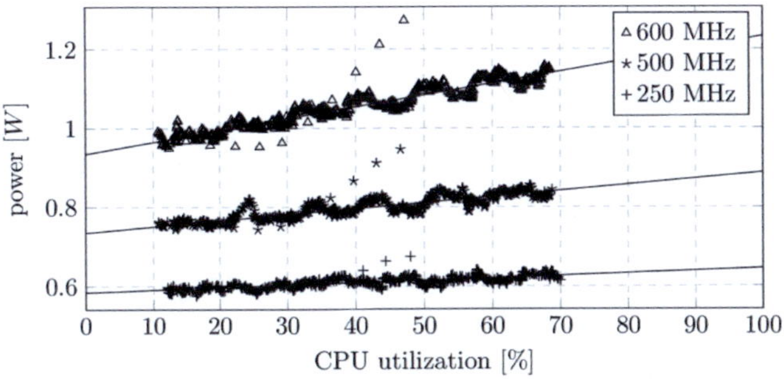

Figure 2: *Power Consumption of GUMSTIX Board depending on CPU Utilization for Different Operating Frequencies.*

The Dummy-ECUs run a Linux operating system with a modified scheduler that combines the scheduling of tasks and the power management of the ECU components (e.g. switching of power states). This allows the implementation of different power management strategies by using so-called power management plans. More details about this model can be found in [1].

By using an additional FPGA, power management techniques like Pretended Networking or Partial Networking can be evaluated where the GUMSTIX board is switched off and the FPGA is monitoring a predefined wake up condition. This might be an external interrupt, a timer, or a certain message on one of the attached communication buses. Each Dummy-ECU features a CAN and Ethernet interface. Furthermore, additional interfaces have been integrated in order to drive external loads.

As shown in figure 3, the Dummy-ECUs are grouped into three domains that are connected through an Ethernet network. The ECUs inside one domain use a CAN network for communication. Messages sent from one domain to another one need to be forwarded by the three main controllers.

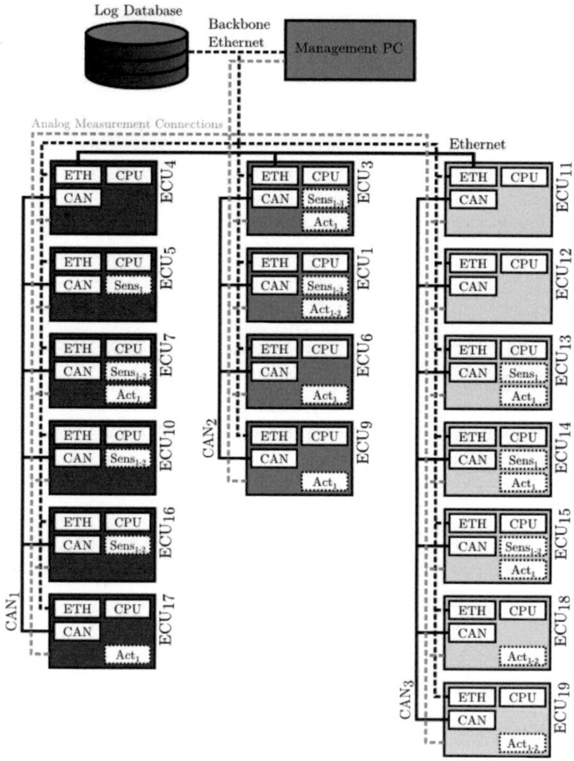

Figure 3: *Schematic Representation of the IT Architecture and Logging Infrastructure.*

All ECUs use the Ethernet network also for transmitting logging information to the central LabVIEW [4] Management PC that is capturing all relevant information into a database. Since the logging information should not have a negative effect on the ECU communication over the Ethernet network, two virtual local area networks (VLANs) are used with different priorities.

As the delay of the communication may have a big impact on the behavior of the e/e architecture we have measured the timings with a small part of the complete system. ECU5 sends a message to ECU1 over one CAN bus 1. It is then forwarded to ECU3 using Ethernet and finally to ECU4 over CAN bus 2. Figure 4 shows the delay of the message from the start of transmission in ECU5 to completion of forwarding or receiving in the following hops. The delay increases with the size of the message with the shape of a step function. The first one is the result of the buffer behavior of the used Linux kernel whereas the other ones have a step each eight byte which is exactly the payload size of a CAN packet.

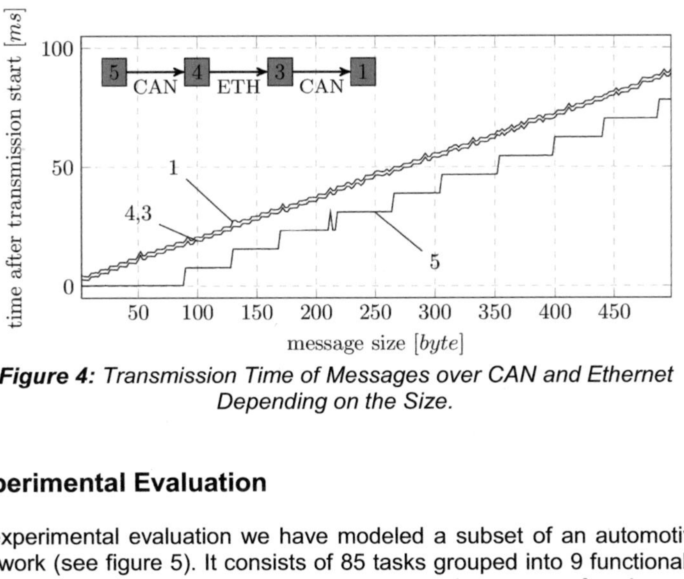

Figure 4: *Transmission Time of Messages over CAN and Ethernet Depending on the Size.*

4. Experimental Evaluation

For the experimental evaluation we have modeled a subset of an automotive functional network (see figure 5). It consists of 85 tasks grouped into 9 functional chains. The functional chains represent various vehicle and customer functions that are needed during different driving situations. Furthermore, we assume a driving cycle where the vehicle is operated in three different modes (see figure 6).

During the first 20 minutes of the driving cycle the vehicle is in the "Driving Mode" where the speed of the vehicle is based on the New European Driving Cycle (NEDC) [5] and all driving related functions need to be switched on. However, there are some functions that are only needed in situations when the vehicle is moving with a speed lower than 30 km/h. An example might be the park distance control that is only needed in slow maneuvering situations and can be turned off for the rest of the time. Furthermore, also other customer functions (e.g. seat heating, audio playback, etc.) are assumed to be switched on and off by the driver during the "Driving Mode".

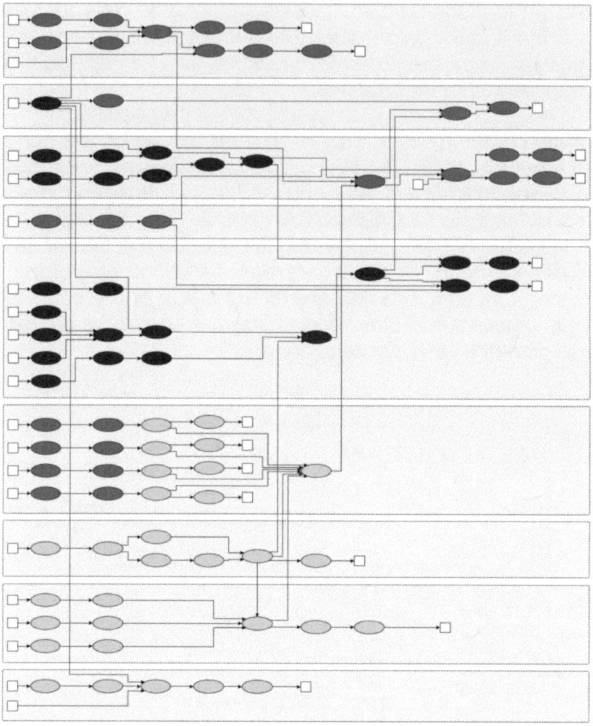

Figure 5: *Exemplary Function Network.*

Afterwards, the vehicle is assumed to be in the "Entertainment Mode" where the vehicle is stopped, the engine is turned off but the driver is still inside the car and uses some of the entertainment and comfort functions of the vehicle. In this mode most of the ECUs can be turned off. The last mode represents the "Parking Mode", where the car is locked and only some security relevant functions need to be turned on (e.g. anti-theft alarm system).

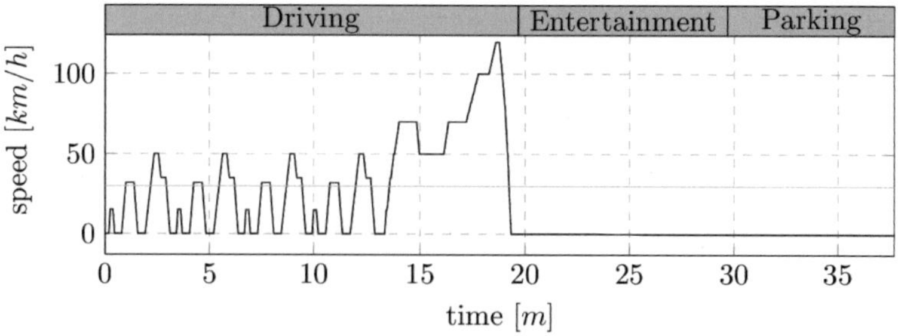

Figure 6: *Vehicle Speed over Time.*

The evaluation was done for three different power management scenarios:

- The first scenario ("No PM") assumes that the e/e architecture doesn't support any power management techniques. All ECUs that are required in a specific driving situation are turned on and are operated in their highest power states. There is not degradation of the system.
- The second scenario ("PM") assumes an advanced power management system. The ECUs can be degraded to more energy-efficient power states or can even be completely turned off whenever it is possible. However, in this scenario we assume that the partitioning of the vehicle functions is not optimal. The vehicle functions that are only needed when the vehicle is moving with a speed lower than 30 km/h are integrated on ECU5 together with other critical functions that can't be switched off. Thus, ECU5 can temporarily be degraded to a lower power state, but cannot be turned off completely.
- The third scenario ("PM + Partitioning") eliminates this limitation by using the advanced power management system together with an optimized function partitioning. Now the full potential of the power management can be enabled by temporarily turning off of ECU5, since only the functions that can be switched off for a vehicle speed over 30km/h are partitioned to that ECU.

Figure 7 illustrates the power consumption over time of the GUMSTIX board of ECU5 for the above mentioned power management scenarios. As can be seen the advanced power management can reduce the power consumption during the "Driving Mode" by degrading the CPU. For the optimized partitioning variant, the whole GUMSTIX board can be temporarily turned off (whenever the vehicle speed is greater than 30 km/h, see figure 6). The functions that are integrated on ECU5 are not needed in the "Entertainment Mode" and "Parking Mode". Thus, the GUMSTIX board is turned off for all three power management scenarios.

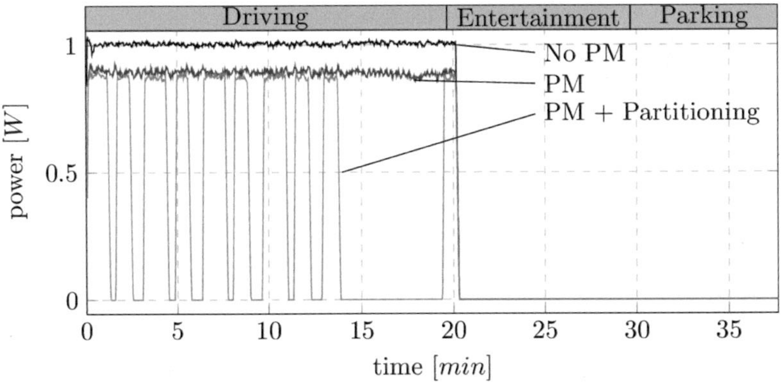

Figure 7: *Power Consumption of the GUMSTIX Board of ECU5.*

In figure 8 the power consumption values of all 17 GUMSTIX boards are accumulated and plotted over time. The effects of the three power management scenarios can be seen. The turning off of ECUs when switching from "Driving Mode" to "Entertainment Mode" and from "Entertainment Mode" to "Parking Mode" is clearly visible.

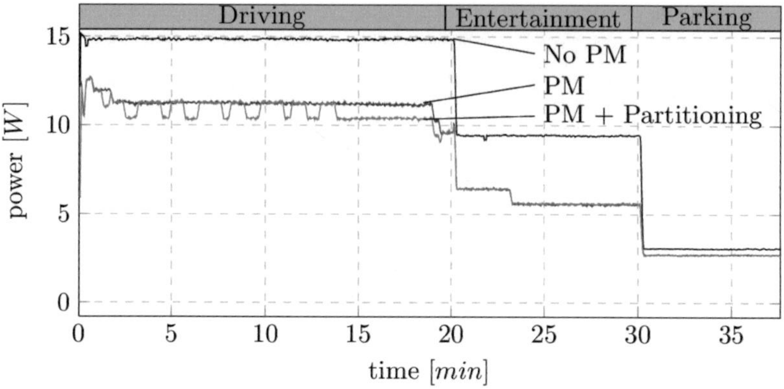

Figure 8: *Accumulated Power Consumption of all 17 GUMSTIX Boards.*

Figure 9 indicates the total power consumption of the whole system. Here not only the power consumption of the GUMSTIX boards is measured, but also of the sensors and actuators. Those sensors and actuators have a significant higher power consumption than the mobile ARM CPU on the GUMSTIX boards and can be turned on and off (demand dependent for all three power management scenarios). The high peak at the beginning is resulting from the seat heating that is used by the driver during the first minutes of the driving cycle.

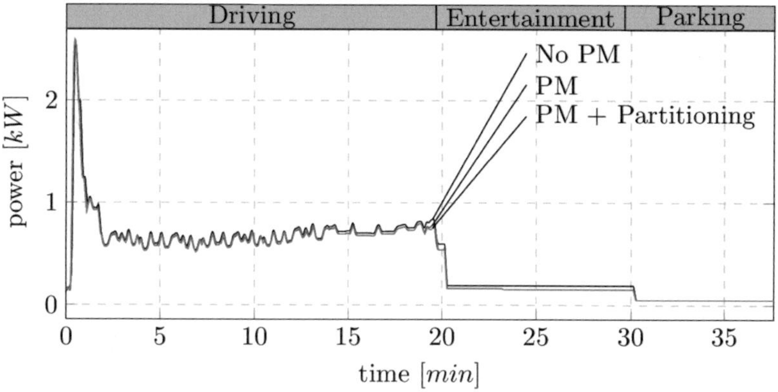

Figure 9: *Accumulated Power Consumption of all Components (GUMSTIX Boards, Transceivers, Sensors, Actuators) of all ECUs.*

5. Conclusion

In this paper a hardware test bench for the evaluation of e/e architectures with respect to energy efficiency was presented. The effectiveness of different power management techniques can be explored and the influence of the function partitioning investigated. Future work includes the evaluation of more advanced functional networks and power management concepts.

Acknowledgments

This work is part of an interdisciplinary research project of the Technical University of Munich and BMW Group Research and Technology. It is organized under the CAR@TUM initiative, where the decades long-standing cooperation between the Technical University of Munich and the BMW Group is intensified and restructured.

Literature

[1] Barthels, A., Ruf, F., Walla, G., Fröschl, J., Michel, H.U., Baumgarten, U.: A model for sequence based power management in cyber physical systems. In: Information and Communication on Technology for the Fight against Global Warming, pp. 87–101. Springer (2011)

[2] Fuchs, M., Scheer, P., Grzemba, A.: Selektiver Teilnetzbetrieb im Fahrzeug: Eine Realisierung für den CAN-Bus und Adaption auf andere Bussysteme. Automotive meets Electronics (2010)

[3] Gumstix Incorporated: Overo Earth COM – Product overview, Website, https://store.gumstix.com/index.php/products/211/

[4] National Instruments Corporation: LabVIEW User Manual, Website, 2003. Available online at http://www.ni.com/pdf/manuals/320999e.pdf

[5] Samuel, S., Austin, L., Morrey, D.: Automotive test drive cycles for emission measurement and real-world emission levels-a review. Proceedings of the Institution of Mechanical Engineers, Journal of Automobile Engineering (2002)

[6] Schmutzler, C., Kruger, A., Schuster, F., Simons, M.: Energy efficiency in automotive networks: Assessment and concepts. In: High Performance Computing and Simulation (HPCS), 2010 International Conference on. IEEE (2010)

[7] Schmutzler, C., Simons, M., Becker, J.: On demand dependent deactivation of automotive ecus. In: Design, Automation Test in Europe Conference Exhibition (DATE), 2012. pp. 69 –74 (march 2012)

[8] Texas Instruments Incorporated: OMAP35x Applications Processor – Technical Reference Manual, Online version, http://ti.com/lit/ug/spruf98y/spruf98y.pdf

[9] Walla, G., Barthels, A., Ruf, F., Dörfel, R., Michel, H.U., Fröschl, J., Sirch, O., Baumgarten, U., Herzog, H.G., Herkersdorf, A., "Framework and Model for the Evaluation of Energy Efficiency of Partitioning Alternatives", "Elektrik/Elektronik in Hybrid- und Elektrofahrzeugen und elektrisches Energiemanagement", Miesbach, Germany, 2012

17 A system-orientated approach towards an integrated module-oriented model-based Electrical/Electronic architecture design

Dipl.-Inform. **Stefan Raue**, Dipl.-Ing. **Markus Conrath**, Dr. rer. nat. **Bernd Hedenetz,** Daimler AG, Sindelfingen, Prof. Dr. rer. nat. **Wolfgang Rosenstiel**, University of Tübingen, Tübingen, Germany

Abstract

In today's electric/electronic (E/E) architecture design for vehicles, model-based methods are used to control the complexity and perform optimisations. Here, a joint consideration of software and hardware extends is necessary. In the early stages of development, the development and modelling in the function- and networking-layer is performed in parallel and very extensive models with complex dependencies between the layers exist. As an additional requirement, the networking hardware components are managed and evolved in modular construction kits. However, e-drive and other innovations will further increase the complexity and have a major impact on the Electrical/Electronic architectures of modern premium cars. Modelling efficiency, the reduction of modelling effort and the improvement of model quality are thus major challenges for the model based E/E architecture development. Hence, a novel concept tackling these issues is presented in this paper. The new reuse concept system construction kit for E/E systems proposes a holistic model of an E/E system embracing the E/E modules. The concept replaces the separate modelling in the various layers and shows an approach to model and manage E/E systems central and across layers. The novel concept thus allows a transparent description of E/E systems as well as the description of the dependencies between E/E systems, functions, and E/E modules. Besides the model quality, the concept improves the advanced E/E architecture design by reusing the modelled E/E systems. Especially in domains with fast changing requirements and constrains, like e-drive, the concept thus provides a rapid and powerful approach for conceptual work and assessment in the E/E architecture development.

1. Introduction

Today, the innovative electric/electronic (E/E) based functions of modern premium vehicles are almost always realised by E/E systems with their networked E/E components (sensors, Electronic Control Units (ECUs), or actuators). The vehicle E/E architecture characterises the interfaces, structure, and interaction of the networked E/E components, power distribution, and wiring harness. An E/E architecture platform includes a family of E/E architectures from which various vehicle model series are derived, such as the S-, E-, and C-class at Mercedes-Benz. Simplified, the development of E/E architecture can be divided into three phases: In the concept phase, the requirements regarding the configuration packages, the vehicle derivatives, the installation of hardware components and the innovations with a strong influence on an E/E architecture are determined and a first E/E architecture concept is designed. In the following detailing phase, the concept capability of all innovations is established and the E/E architecture concept is adjusted accordingly. In the final serial phase, further adjustments to the E/E architecture concept are possible which can result from competition, laws, optimisations or model facelifts.

E/E architectures are subject to a heavy influence from hardware component modularisation and thus modular construction kits. The hardware modularisation is applied by OEMs (Original Equipment Manufacturers) to reuse once developed hardware components in several vehicle model series. In general, modularisation enables to design modular product architectures [1] and is in industry differently performed [2]. In the automotive industry, an assembly orientated modularisation is dominating. Here, a module is formed by those hardware components that are in the production installed together into the vehicle. Concerning the E/E, an E/E module is composed of sensors, ECUs and actuators that belong to an assembly unit, e.g. door. A further heavy influence on E/E architectures originates from the distributed functions of the function network. These and the realising hardware are organised in E/E systems. The E/E architecture concept is the source for the E/E system development in an integrated and seamless development process [3]. The function part responsible for the realisation on the hardware component is termed function contribution (FC). Usually, various functions have a FC assigned to one hardware component and they all are implemented by AUTOSAR software components (SWCs).

In the concept phase, models of the E/E architecture contain different layers with complex dependencies relating them. Some of these dependencies can be modelled increasing the transparency in the development, like the mapping between FCs and hardware of an E/E module. Although further dependencies between hardware of the E/E modules arising from distributed functions of the function network exist, the layers are modelled separately and E/E systems as holistic structuring element are not regarded. Thus, the dependencies between the E/E modules are intransparent and so it is still a challenge to model all aspects of an E/E system. Additionally, a modelling efficiency increasing reuse of modelled extents is just available for E/E modules without inclusion of the functions. At the same time, a well-known trend in the automotive industry is the further increase of the function range and therefore E/E complexity [4]. These facts motivate to extent the E/E architecture design methodology by a new concept, which is presented in this paper. The new concept of a system construction kit is intended to cope with the increasing complexity of functions on the one hand and with the hardware modularisation on the other hand. Moreover, we see the capture of the high degree of interaction between different E/E systems, their evaluation, their optimisation, and the generation of an overall picture as a new and necessary task in an E/E architecture development. The novel system-oriented design methodology with the system construction kit thus has the goal to replace today's parallel development of functions and hardware by an integrated approach.

2. Previous Work

It was very early decided by Daimler AG to perform the concept phase of an E/E architecture family development using a model-based approach [4]. From the today's perspective, a model-based E/E architecture design is state-of-the-art and performed by most competitors and suppliers. The distribution of modelling objects of an example E/E architecture platform model with an extent of circa one million model objects is shown in Figure 1a. The major share of the model objects has the function network, which can be also denoted as logical architecture [5]. Apart from the function network, the hardware networking (including wiring harness and power distribution) and topology is modelled (Figure 1b). Model objects of different layers are related by mappings [4]. E/E systems include both hardware components and functions of the function network and are currently not modelled as structure element in the E/E architecture modelling. In the last years, the E/E architecture design methodology was significantly enhanced allowing a modelling of E/E modules of the module strategy [6]. The modules can be integrated in the model-based E/E architecture design [7] using a product line approach [8]. In the modular construction kit, the E/E modules can be modelled, evolved, and reused in different E/E architecture models (Figures 1c and 2). However, the function network is modelled separately from the module construction kit and is maintained in parallel in the different E/E architecture models (Figure 2), which causes extra work in the modelling. Additionally, the assignment of functions and E/E modules to E/E systems is not captured holistically.

In the field of the automotive software development, the reuse of already developed results was recognised as very important and necessary to increase the productivity [9] and a reuse approach was suggested [10]. In the field of model-based E/E architecture development, reuse was firstly performed for hardware of E/E modules, as described above. Recently, a tier-1 supplier reported of an extended reuse including the layer of software architecture and logical layer to reuse model parts within the

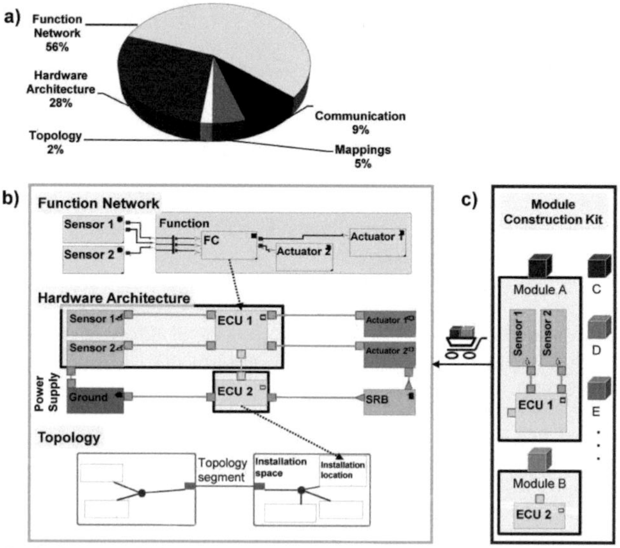

Figure 1: *Model based E/E architecture design. a) Distribution of modelling objects. b) Modelling in layers. c) E/E systems in the modelling levels. d) Module construction kit according to [6].*

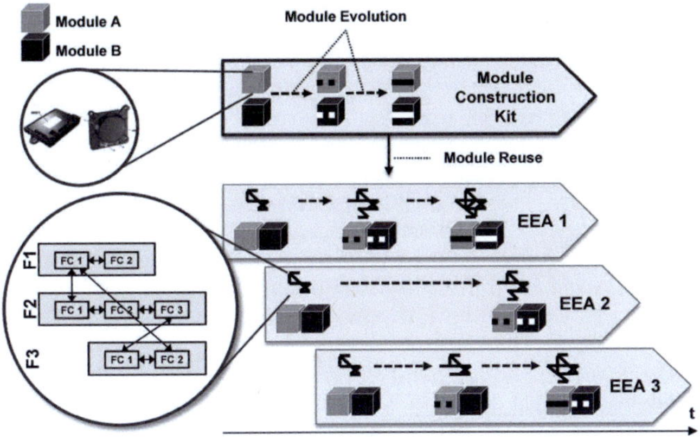

Figure 2: *Hardware module construction kit. EEA: E/E architecture.*

development of the serial phase [11]. For the OEM concept phase, however, a reuse approach covering an E/E system as whole including the function network layer is still missing. To involve the topology layer, however, is not necessary since it is primarily just used for wiring harness length calculations and varies to a large amount between the vehicle series.

Thus, we regard the contribution of this paper as threefold. First, a novel concept for a central modelling of E/E systems across layers is presented which replaces the parallel modelling in various layers. Second, we show how this approach allows a transparent description of E/E systems as well as a description of the dependencies between E/E systems, functions, and E/E modules is proposed. Last, experiences using this approach in the productive model-based E/E architecture development are shared.

3. Use cases and Requirements

Extension of the Module Construction Kit: An E/E system includes several hardware components by distributed functions. These hardware components can belong to different E/E modules. Additionally, the E/E modules may have variants arising e.g. out of the scaling. Accordingly, a basis variant of the E/E module, for example, could contain one ECU and one sensor while the premium variant could have the ECU with 3 sensors. Currently, a methodology is used which allows to model each variant separately and so the same ECU, for instance, is modelled several times. Instead of these several 100% E/E module models, a 150% representation of the E/E modules is necessary for the novel concept while before just several 100% E/E module representations were sufficient and artefacts were duplicated.

Management and structure: An E/E system can have variants and also require- or exclude-relationships to other E/E systems. A system construction kit must be able to manage and describe these relationships. Additionally, the functions of E/E systems should be considered system construction kit concept. The distribution of functions and the resulting functional coupling of the E/E modules are currently not considered by the assembly-oriented modularisation of the hardware. From E/E architecture perspective, three types of E/E module couplings exist, which can be distinguished and classified as follows (Figure 3):

- *Distributed functions:* E/E modules receive a direct and transparent coupling if the functions of an E/E system are distributed to hardware components of different E/E modules. This distribution of the functions limits exchangeability of the E/E modules. Regarding the hardware layer, two E/E modules are exchangeable if they can be easily integrated by having the same bus connections (e.g. 1xCAN and 1xFlexray). If the functional layer is also regarded, the two E/E modules are not easily exchangeable because the FCs of the functions allocated on the two E/E models must match in their functionality and their interfaces.
- *Function reuse and implementation alternatives:* E/E modules receive an indirect and transparent coupling if FCs are reused in different E/E modules or E/E module variants. In Figure 3b, FC1 is reused denoted by FC1' and FC1", respectively. For these reused FCs, implementation alternatives arise if they are developed by different suppliers
- *Commonalities in implementation:* If the implementation of FCs by SWCs is additionally regarded, further E/E module couplings can arise. SWCs are either clearly assignable to FCs or not. In a clear assignment, several SWC are assigned to one FC resulting in an 1-n-relationship. However, in case of a central coordinator SWC, for example, this SWC is assigned to several FCs which results in an indirect and non-transparent module coupling. As shown in Figure 3c, FC3 of function 3 is implemented by the SWCs B, C, and E and FC1 of function 4 is implemented by the SWCs A, C, and D. Thus, both function 3 and 4 share SWC C as communality in their implementation.
- *Hardware-driven E/E systems:* E/E modules receive a transparent but non-relevant coupling by hardware-driven E/E systems. Usually, they are in small number but distributed over many E/E modules. These E/E systems (e.g. diagnostics systems) just add functionality because the hardware is present.

Evolution: To describe the evolution of E/E systems and extend the methodology, past E/E system evolutions were analysed. From the E/E architecture perspective, the following typical evolution use cases for E/E systems distributed across multiple modules were derived:

- *New and die:* This use case is the trivial use case which characterises the emergence and disappearance of FCs on existing E/E modules.
- *Move:* This use case describes the change where, especially in the context of optimisations, FCs are shifted to other already existing E/E modules.
- *Split:* A new variant of the E/E module might adopt only a part of the E/E system function range compared to the existing E/E module. Such a function range reduction can be realised, for instance, for a new module variant intended for the compact segment.

Reuse: Currently, E/E modules contained in the module construction kit can be reused in different E/E architectures. However, the function network must additionally be modelled separately for all E/E architectures the hardware is reused in. This modelling creates overhead which should be reduced by a new concept. Thus, E/E systems which are once described in the system construction kit should be reused as whole including the E/E modules and the function net.

Information generation: Since a new concept should cover more modelling extends, new views are necessary to check the model and create transparency, e.g. a view to show which E/E systems contain which E/E modules.

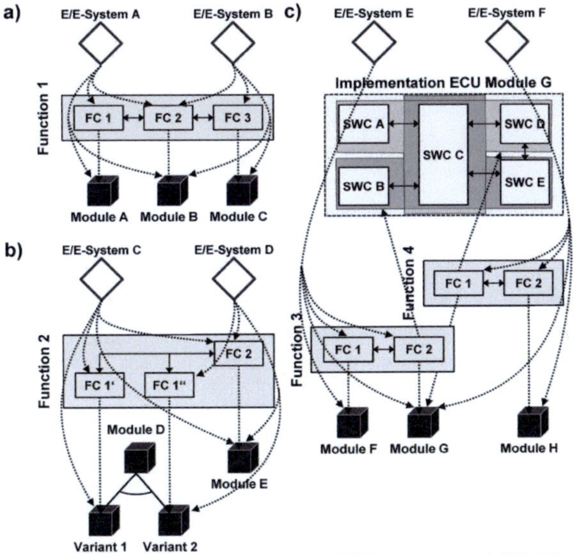

Figure 3: *E/E module coupling and linkage to E/E systems. a) Distributed functions. b) Function reuse and implementation alternatives. c) Communalities in implementation.*

4. Proposed approach: System Construction Kit

In this section, the novel concept for a system construction kit is introduced which provides a holistic model of an E/E system embracing all necessary layers, extending the module construction kit and allowing to structure, reuse and evolve the E/E systems (Figure 4).

Extension of the Module Construction Kit: To extent previous approaches, the information describing the assignment of hardware components to E/E modules is transferred to a separate layer. This decoupling allows indicating hardware reused in different E/E module variants. Additionally, relationships can be created helping to describe variants that exclude or extent another. Together, creating 150% E/E modules models is allowed (Figure 4a).

Management and structure: To model E/E systems, an E/E system model as part of the system construction kit is created (Figure 4a) with the following steps. First, a central E/E system list containing the E/E systems of a whole E/E platform is created.

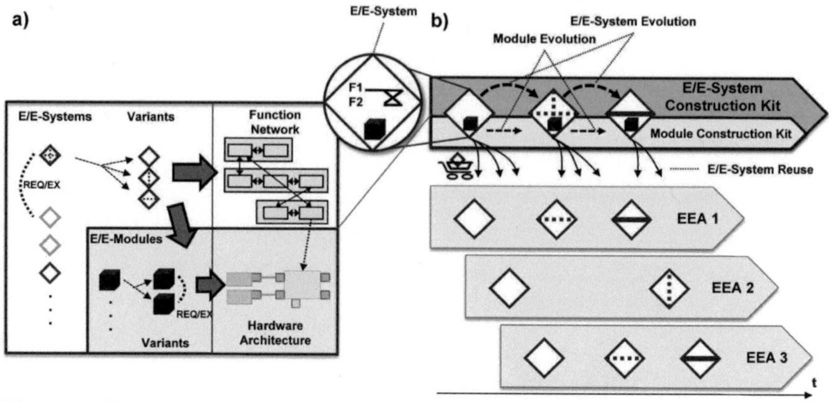

Figure 4: *System construction kit concept. a) Management and structure. b) reuse and evolution. EEA: E/E architecture.*

Furthermore, the E/E system list is enriched by requires- and exclude-relationships between the E/E systems. Additionally, the E/E system list is completed by extensive innovations which will be covered by an own E/E system instead extending an already existing E/E system. Second, this central list is used as anchor to model the linkage to all elements which belong to an E/E system but are distributed to several layers. Functions of the function net and E/E modules of the module construction kit can be linked to E/E systems according to the described use cases above. Third, the so described E/E systems form the base for the variant management. For a vehicle variant analysis, just a selection which E/E system is contained in the vehicle variant is necessary since all extends of the E/E system are described by the linkage. However, a different mechanism is used for hardware-driven E/E systems. Here, FCs and hardware are linked to E/E systems, which are just in the vehicle when the hardware already exists. Thus, a further refinement step must be added to check which hardware exists and add the hardware-driven E/E systems in this case.

Evolution: Given the range, the complexity and the tight weaving of E/E systems, functions and E/E modules, a model-based evolution is difficult. Here, a deep understanding of the functions is necessary and special issues are complex, bound to expert knowledge and require a tedious clearing. A supporting methodology utilizing a conceptual function architecture layer as further abstraction allowing design decisions was proposed recently [12]. For modelling the design decisions, however, the central execution of the described practical use cases is enabled by the system construction kit (Figure 4b). This is possible since all necessary information of E/E systems is managed centrally. Furthermore, the changes can be illustrated and an automatic evolution of the already existing reuses of E/E systems in different E/E architectures is allowed. Even complex evolution scenarios can be performed by creating new E/E system variants or E/E module variants with according linkage to function network and hardware architecture, respectively.

Reuse: A reuse in different E/E architecture models is advantageous since otherwise the same extents have to be modelled several times (Figure 2). Due to the central modelling in the system construction kit, a reuse of whole E/E systems is proposed

(Figure 4b). This central modelling and reuse replaces the parallel maintenance of the function network in the different E/E architecture models.

Information generation: The concept also allows creating general views for the whole vehicle, e.g. tables to show which E/E systems are distributed to which module variants or which E/E systems contain which FCs.

5. Experiences

The evaluation of the concept was done in a chassis/driving assistance E/E architecture development project at Daimler AG. Destination for an implementation of the above described concept is the modelling tool [13], which is used by Daimler AG to model the E/E architectures in the early development phase and is involved in the overall vehicle development process. In the tool, the different layers are in use where the functions are modelled in the form of a function network and the E/E modules on the hardware layer. An extension of the tool with the system construction kit concept is possible by plug-ins and metrics. To ensure operations, this extension should not restrict the current methods and processes but merely expand them. Thus, the tool was first enriched for the most parts of the concept described above by a prototypical implementation utilising a copy of productive models for testing. Here, the ADAS module model extents were in focus. Subsequently, the management and structure - parts of the concept were used for the next generation of Daimler AG concept E/E architecture chassis/driving assistance productive models which cover the next generation of Mercedes-Benz S-class.

Extension of the Module Construction Kit: To manage the 150% module models, the E/E system models, and the relations, a linkage of hardware and software artifacts is necessary. This is possible in the requirements layer and the costumer feature layer. The latter is designed to model all orderable features of the product and its variance. With the modelling of E/E modules, the task of this layer is expanded and no longer limited to customer features. The linkage was established by a mapping to the corresponding modelling objects in the hardware layer. Using this mechanism, the creation of new E/E module revisions was found to be very straightforward. Since the 150% modelling, an extensive selective copying of model extends can be omitted. Additionally, the model size and the modelling effort could be reduced because just one instance of ECU, sensors and actuators in the library is necessary. As result, the number of software mappings to the different instances could be reduced and thus the clarity of the model increased. However, the ECU instance increases over time more and more. It should thus be decided in some point to make a branch of the E/E module as new separate version, for example after a revolutionary change.

Management and structure: Analogous to the E/E modules, also the customer feature layer was extended in its sense to model the technical E/E systems information like described above. This creates a holistic model of an E/E system across all layers containing all involved artefacts. However, modelling a whole domain with this approach requires an initial effort to model the relations and especially to link the numerous corresponding FCs. Nonetheless, now a mechanism is provided to capture all transparent E/E module couplings. The intransparent E/E module coupling, however, requires a further layer for their description wherefore a conceptual function architecture layer is proposed [12]. The layer used for the description of the E/E systems has an association to the variant management allowing including or excluding costumer features in the defined E/E architecture variants. This results in a simpler variant configuration possibility leading to a powerful systematic E/E system variability management. Apart from this, the model transparency and the modelling quality increased since the E/E architects were relieved of concerns which were managed manually by the E/E architects before.

Evolution: This method also improved the flexibility. Several possibilities opened up to model new extents. It was thus, for instance, possible to just create a hardware model revision ignoring the functions at first. This could be done later by just adjusting the linkage accordingly. At the same time, E/E modules not contained in the considered platform can be kept as a library by just removing the linkage to the E/E systems. Examining different concepts also prove to be effortless since just the linkage must be adapted to compare concepts with, for instance, three versus four cameras. For assessment of the evolution, a conceptual function architecture layer is proposed [12] while a description of the modelling of the evolution including an automation of the above described use cases will be concern of later publications.

Reuse: The E/E systems, which are once described as a holistic model using this mechanism, enable the reuse of E/E systems throughout E/E architecture models since all the artefacts belonging to this E/E system are involved. For the reuse, a relatively new modelling tool mechanism is exploited, which so far had to be self-implemented for the reuse of E/E modules. Now, the modelling tool allows defining a container for various model objects manually and reusing this container in E/E architectures. Accordingly, a container is defined in the system construction kit for model objects of an E/E system and so the E/E system is reused in the relevant E/E architecture. Additionally, for automating the tedious manual definition of this container, a novel metric was developed which allows a dynamic creation of this container for a given E/E system model. This so performed dynamic reuse was found to increase the modelling efficiency.

6. Conclusion

Today's E/E architecture development is realised by a parallel modelling of specific aspects such as functions, networking, and wiring harness at different layers. Our current models attain a very large model range, a high degree of complexity and a large number of dependencies. Regarding to the trends in the automotive industry, they will continue to grow strongly - particularly at the functional layer. Thus, important aspects are the development and reuse of model objects. As approach, the system construction kit and thus a new E/E architecture modelling concept was presented. Advantages are the transparent description and methodological development of dependencies between E/E systems, functions, function contributions, SWCs and E/E modules. By the central modelling of E/E systems, which can be reused in the different E/E architectures, the system construction kit concept promises to manage the complexity, an increased modelling efficiency, and a support for the E/E architect. In total, this integrated approach for a holistic consideration of E/E systems proves to be rapid and powerful for conceptual work and assessment of E/E architecture development.

References

[1] K.T. Ulrich, and W.P. Seering, “Function sharing in mechanical design”, Artificial intelligence in engineering design 2, pp. 185-213, 1992

[2] C. Florent, “Strategic perspectives on modularity”, Dynamics of industry and innovation: organizations, networks and systems, The DRUID tenth anniversary summer conference, pp. 27-29, 2005

[3] S. Schmerler, T. Ringler, B. Hedenetz, U. Grüner, F. Wohlgemuth, C. Dziobek, P. Lohrmann, „Mit AUTOSAR zu einem integrierten und durchgängigen Entwicklungsprozess“, 15. Internationaler Kongress Elektronik im Kfz, Baden-Baden, VDI-Verlag GmbH Düsseldorf, October 2011

[4] R. Belschner, J. Frees, and M. Mroßko, „Gesamtheitlicher Entwicklungsansatz fur Entwurf, Dokumentation und Bewertung von E/E-Architekturen“, 12. Internationaler Kongress Elektronik im Kraftfahrzeug, Baden-Baden, VDI-Berichte Nr. 1907, VDI-Verlag GmbH Düsseldorf, 2005

[5] M. von der Beeck, „Development of logical and technical architectures for automotive systems”, Software & Systems Modeling, vol. 6(2), pp. 205-219, 2007

[6] M. Jaensch, P. Prehl, G. Schwefer, K.D. Müller-Glaser, and Markus Conrath, „Model-Based Design of E/E Architectures 2.0“, ATZelektronik worldwide eMagazine, vol.6(4), pp. 16-21, 2011

[7] M. Jaensch, M. Conrath, B. Hedenetz, and K.D. Müller-Glaser, “Integration of Electrical/Electronic-Relevant Modules in the Model-Based Design of Electrical/Electronic-Architectures”, 1. Energy Efficient Vehicle Conference (EEVC), pp. 84–92, June 2011

[8] M. Jaensch, B. Hedenetz, M. Conrath, and K.D. Müller-Glaser, „Transfer von Prozessen des Software-Produktlinien Engineering in die Elektrik/Elektronik-Architekturentwicklung von Fahrzeugen“, 8. Workshop Automotive Software Engineering, pp. 497-502, September 2010

[9] K. Grimm, „Software technology in an automotive company: major challenges”, Proceedings of the 25th international conference on Software Engineering, IEEE Computer Society, pp. 498-503, May 2003

[10] B. Hardung, T. Kölzow, and A. Krüger, "Reuse of software in distributed embedded automotive systems", Proceedings of the 4th ACM international conference on Embedded software. ACM, 2004

[11] M. Kardos, “Product line management – a key for reuse”, PREEvision User Day 2012, November 2012

[12] S. Raue, M. Conrath, B. Hedenetz, W. Rosenstiel, „Assessing the evolution of E/E hardware modules with conceptual function architectures“, Advanced Microsystems for Automotive Applications 2014, June 2014

[13] T. Ringler, M. Simons, and R. Beck, “Reifegradsteigerung durch methodischen Architekturentwurf mit dem E/E-Konzeptwerkzeug”, 13. Internationaler Kongress Elektronik im Kraftfahrzeug, Baden-Baden, VDI-Berichte Nr. 2000, VDI-Verlag GmbH Düsseldorf, 2007

The Authors

Univ.-Prof. Dr.-Ing. Bernard Bäker
Dresden Institute of Automobile Engineering – IAD
Technische Universität Dresden, Dresden, Germany

Prof. Melody Baglione
The Cooper Union for the Advancement of Science, New York, USA

M.Sc. Andreas Barthels
Technische Universität München, Munich, Germany

Dipl.-Inform. Tobias Bär
FZI Research Center for Information Technology at the Karlsruhe Institute of Technology, Karlsruhe, Germany

Dipl.-Ing. Salim Chaker
ITI GmbH, Dresden, Germany

Mark J. Duty, B.S.E.E.T
Chrysler LLC, Michigan, USA

Dipl.-Ing. Markus Conrath
Daimler AG, Sindelfingen, Germany

Prof. Yasuhiro Daisho
WASEDA University, Tokyo, Japan

Dr.-Ing. Marjam Eghtessad
Institute of Automotive Engineering Braunschweig University of Technology, Braunschweig, Germany

Dipl.-Ing. Rainer Falsett
Daimler AG, Böblingen, Germany

Dr.-Ing. Thomas Forchert
Celono, Berlin, Germany

Dipl.-Ing. Joachim Fröschl
BMW Group, Munich, Germany

Dipl.-Ing. Johannes Fuchs
Graz University of Technology, Graz, Austria

B.Sc. Dirk Gabriel
Technische Universität München, Munich, Germany

Dipl.-Ing. Mykhailo Gurtovyi
Kharkiv National University of Radio Electronics, Kharkiv, Ukraine

Dr. rer. nat. Bernd Hedenetz
Daimler AG, Sindelfingen, Germany

Dipl.-Ing. Martin Heinrich
EA Systems Dresden GmbH, Dresden, Germany

Dr.-Ing. Wolfgang Herdeg
Dr. Ing. h. c. F. Porsche AG, Weissach, Germany

Prof. Dr. Andreas Herkersdorf
Technische Universität München, Munich, Germany

Prof. Dr. Dr. Gerhard Hettich
EAST-Consulting, Dietenhofen, Germany

Dr.-Ing. Georg Immel
BMW Group, Munich, Germany

Prof. PhD.eng. Yushi Kamiya
WASEDA University, Tokyo, Japan

Dr.-Ing. Steffen Kehl
Dr. Ing. h. c. F. Porsche AG, Weissach, Germany

Dipl.-Ing. Christian Kehrer
ITI GmbH, Dresden, Germany

Dipl.-Ing. Mario Krumnow
Technische Universität Dresden, Dresden, Germany

Prof. Dr. Olexandr Kuzemin
Kharkiv National University of Radio Electronics, Kharkiv, Ukraine

Prof. Dr.-Ing. Ferit Küçükay
Institute of Automotive Engineering (IAE)
Braunschweig University of Technology,
Braunschweig, Germany

Dipl.-Ing. Per Lewerenz
Technische Universität Dresden,
Dresden, Germany

Prof. Dr. Lennart Löfdahl
Applied Mechanics, Chalmers
University of Technology, Gothenborg,
Sweden

Dipl.-Ing. Sven Löwenberg
Technische Universität Dresden,
Dresden, Germany

Dipl.-Ing. Alexander Mai
BMW Group, Munich, Germany

Dipl.-Phys. Hans-Ulrich Michel
BMW Forschung und Technik GmbH,
Munich, Germany

Dr.-Ing. Dennis Nienhüser
FZI Research Center for Information
Technology at the Karlsruhe Institute
of Technology, Karlsruhe, Germany

Dipl.-Ing. Ronny Petersohn
Technische Universität Dresden,
Dresden, Germany

Prof. Cand. Sc. Viktor Pismenetskyi
Kharkiv National University of Radio
Electronics, Kharkiv, Ukraine

Dr. rer. nat. Hartmut Pröbstle
BMW Group, Munich, Germany

Dipl.-Inform. Stefan Raue
Daimler AG, Sindelfingen, Germany

Dipl.-Ing. Robert Richter
Technische Universität Dresden,
Dresden, Germany

Prof. Dr. rer. nat. Wolfgang Rosenstiel
University of Tübingen, Tübingen, Germany

Dipl.-Ing. Florian Ruf
Technische Universität München,
Munich, Germany

B.eng. Hiroki Sato
WASEDA University, Tokyo, Japan

Dipl.-Ing. Victor Schäfer
Dr. Ing. h. c. F. Porsche AG, Weissach,
Germany

Dipl.-Ing.Heiner Schmeck
CTC cartech company GmbH, Boeblingen,
Germany

Dr. techn. Bernhard Schweighofer
Graz University of Technology, Graz, Austria

Dipl.-Ing. Torsten Schubert
Technische Universität Dresden, Dresden,
Germany

M.Sc. Mark Schudeleit
Institute of Automotive Engineering (IAE)
Braunschweig University of Technology,
Braunschweig, Germany

Dipl.-Ing. Ottmar Sirch
BMW Group, Munich, Germany

Prof. Dr. Mykola Slipchenko
Kharkiv National University of Radio
Electronics, Kharkiv, Ukraine

Dipl.-Ing. Stephan Uebel
Technische Universität Dresden, Dresden,
Germany

Dipl.-Ing. Andreas Unger
IAM GmbH, Dresden, Germany

Dipl.-Ing. René Unger
EA Systems Dresden GmbH, Dresden,
Germany

Dipl.-Ing. Gregor Walla
Technische Universität München, Munich,
Germany

Dr. techn. Hannes Wegleiter
Institute of Electrical Measurement and
Measurement Signal Processing,
Graz University of Technology, Graz, Austria

M.Sc. Michael Winter
Technische Universität München, Munich,
Germany

M.eng. Wei-Hsiang Yang
WASEDA University, Tokyo, Japan

Prof. Dr.-Ing. J. Marius Zöllner
FZI Research Center for Information
Technology at the Karlsruhe Institute
of Technology, Karlsruhe, Germany

Diagnose in mechatronischen Fahrzeugsystemen VIII

Neue Verfahren für Test, Prüfung und Diagnose von E/E-Systemen im Kfz

ISBN 978-3-944331-59-1, 39,80 EUR [D]

Das Thema *Diagnose*, in Verbindung mit dem Test und der Prüfung technischer Systeme, kristallisiert sich immer mehr als bereichsübergreifende Kernkompetenz der Fahrzeugentwicklung heraus. Insbesondere Sicherheit und Zuverlässigkeit im Zusammenhang mit neuen domänenübergreifenden vernetzten Funktionen sowie hochgradig vernetzte hybride Antriebs- und Assistenzsysteme stellen an die *Diagnose* immer neue Herausforderungen. Remotediagnose, „End to End"-Diagnosemethoden bei domänenübergreifenden Systemen, neue Standardisierungen wie OTX, ODX, WWH-OBD, wissensbasierte Diagnoseverfahren und neue Fehlererkennungsmechanismen in Steuergeräten prägen dabei die Zukunft der Fahrzeugdiagnose. Anhand aktueller Entwicklungsberichte, Systemvorstellungen und Praxisbeiträge werden in diesem Themenband der aktuelle Stand und die Trends für F&E im Themenfeld Diagnose, Test und Prüfung von Elektroniksystemen im Kraftfahrzeug vorgestellt.

Inhalt: Diagnose 4.0 - Herausforderungen durch zunehmende Vernetzung zwischen Fahrzeug, Kunde und Umwelt – Demonstration of Secure On-Board Diagnostics – Quality of Service in Ethernet-basierten Fahrzeugnetzwerken und deren Auswirkungen auf die Diagnose – Erfahrungsbericht Remote Diagnose – Diagnose der nächsten Generation - Prüfabläufe mit OTX beherrschen – Die OTX Programmierung in der Praxis bei stark variierenden Anforderungen der Anwender – OTX programmierbarer Tester für Inspektionen im Feld (OpTIF) – Visualisierung von Diagnoseprozessen – Fehlererkennung für ein automatisiertes Getriebe in schweren Nutzfahrzeugen – Softwarebasiertes Fehlermanagement in elektronischen Steuergeräten – Methodik und Toolkette für den Diagnose-Designentwurf vernetzter Systeme im Fahrzeug mit Hilfe von PREEvision – Modellbasierte Erzeugung von Diagnosewissen von E/E-Architekturen mit PREEvision, u. a.

Die Herausgeber: Dipl.-Ing. Andreas Unger, Lehrstuhl Fahrzeugmechatronik, Technische Universität Dresden; IAM GmbH, Dresden Univ.-Prof. Dr.-Ing. Bernard Bäker, Institut für Automobiltechnik Dresden – IAD, Lehrstuhl Fahrzeugmechatronik, Technische Universität Dresden

TUD*press*
Verlag der Wissenschaften GmbH
Bergstr. 70 | D-01069 Dresden

Tel.: +49 (351) 47 96 97 20
Fax: +49 (351) 47 96 08 19
mail@tudpress.de